AF817736

INDIA'S NUCLEAR ENERGY PROGRAMME

Challenges, Prospects and Public Concerns

INDIA'S NUCLEAR ENERGY PROGRAMME

Challenges, Prospects and Public Concerns

Editors

Arvind Gupta

K.D. Kapur

G. Balachandran

Kapil Patil

Foreword by

Anil Kakodkar

INDIAN PUGWASH SOCIETY

PENTAGON PRESS

India's Nuclear Energy Programme: Challenges, Prospects and Public Concerns
Arvind Gupta, K.D. Kapur, G. Balachandran, Kapil Patil (Eds.)

First Published in 2014

Copyright © Indian Pugwash Society, New Delhi

ISBN 978-81-8274-781-4

All rights reserved. No part of this publication may be reproduced, stored in a retrieval system, or transmitted, in any form or by any means, electronic, mechanical, photocopying, recording, or otherwise, without first obtaining written permission of the copyright owner.

Disclaimer: The views expressed in this book are those of the authors and do not necessarily reflect those of the Indian Pugwash Society.

Published by
PENTAGON PRESS
206, Peacock Lane, Shahpur Jat,
New Delhi-110049
Phones: 011-64706243, 26491568
Telefax: 011-26490600
email: rajan@pentagonpress.in
website: www.pentagonpress.in

Branch:
Prime Arcade
Office #11
1154 Saifee Street
Opp. M.G.Road, Camp
Pune-411001
Email: pentagonpresspune@gmail.com

In association with
Indian Pugwash Society
1-Development Enclave
No.1 - Development Enclave
Rao Tula Ram Marg,
New Delhi, 110010
Phone: 91-11-2671 7983 - extn 7015, 7015
Website: www.pugwashindia.org

Printed at Aegean Offset Printers, Greater Noida, U.P.

Contents

Foreword

This report on "**India's Nuclear Energy Programme: Challenges, Prospects and Public Concerns**" prepared by Indian Pugwash Society is an important contribution to the discourse on nuclear energy in India and one must begin with complimenting the editors, contributors and the Indian Pugwash Society for this timely effort.

We are all aware of the strong dependence of the Human Development Index (HDI) on per capita electricity use up to a threshold of about 5000 kWh/capita. Given our very low per capita electricity consumption and consequently a fairly low rank in terms of HDI as well as our large population and rapidly growing economy; the need for additional electricity, to reach a given standard of living, is the highest for India. There is however inadequate recognition of the energy crisis that we as a country are heading towards. Studies by Planning Commission of India and others suggest that energy import in India is likely to increase from present 30% to around 60% by the year 2032. As a consequence the energy import bill is likely to rise at GAGR of 18% and 10% for oil and coal respectively. This would take the pay out on this account to Rs.200, 000 billion and Rs.7000 billion respectively. Clearly, such a scenario is not sustainable both economically as well as in terms of stability of global environment.

Resolution of this emerging crisis lies in pursuing strategies that enable meeting our energy needs through resources within the country. To be able to do this on a scale commensurate with our long term needs, which would be around an order of magnitude larger than the present demand, recourse to thorium and solar energy is inevitable. Both these sources also enable us to move away from fossil energy that adds to the threat of global warming. The key question however is how soon the share of these two energy forms can become significant enough to be able to tide over emerging large scale dependence on energy imports with potentially volatile prices.

This would require a holistic national strategy and program implementation, with active involvement of all stakeholders, at a rate significantly faster than general power development in the country. This would also need development/deployment of specific new technologies appropriate to our conditions. Need to depend on Thorium and hence taking recourse to nuclear recycle, ensuring strategic autonomy, high dust load, water scarcity in areas where solar insolation is the highest etc. are some of the India specific challenges that need to be recognised in the new paradigm that should emerge in the context of nuclear and solar energy. The report comprehensively discusses the broader national and international scenario under which the nuclear power development in the country would need to take place.

Damage to nuclear reactors at Fukushima caused by the unprecedented tsunami and large scale displacement of people as a result has clearly been a setback to nuclear energy development worldwide including in our country. The report covers the way different countries have responded to this event in fair detail. Clearly the nuclear industry would learn lessons and make appropriate corrections and emerge stronger as had been the case after Chernobyl. While doing so, I believe, it is also necessary to bring in a balance by eliminating the significant psychosomatic consequences that have come to light as a result of perhaps avoidable over reaction in terms of excessive displacement of people.

It is worth noting that Japan while rethinking their future policies on energy mix has retained nuclear energy option. Every country has to make independent choices based on their needs, available resources, economics and energy security considerations. Clearly in Indian context we should have done better in terms of greater mining efficiency as well as energy use efficiency with Indian coal characterised by high ash content. We need to remain focused on rapid transition to Thorium utilisation at a level commensurate with long term Indian energy needs in case of nuclear energy as well as on large capacity solar plants to be located on barren uncultivable tracts of land away from population in case of solar energy.

It seems to me that the following four broad areas need more serious focussed attention as far as nuclear energy development is concerned;

1. Greater neighbourhood engagement to a level that NIMBY (not in my backyard) syndrome gives way to a strong neighbourhood

preference for locating nuclear power plant in spite of negative campaign by vested interests,

2. Elimination of ambiguity in terms of economic impact on nuclear power costs arising out of our legal framework particularly relating to civil nuclear liability,
3. Near 100% domestic value addition for power projects set up through international co-operation so that the cost of such plants compares favourably with domestic nuclear plants as well as other modes of electricity production. It may also be desirable to facilitate discovery of prices through a transparent competitive bidding process for a new project without diluting existing commitments and
4. Accelerated implementation of second stage (PFBR and follow on plants based on fast reactors) programme along with associated nuclear fuel recycle plants as well as Advanced Heavy Water Reactor (AHWR) that takes safety of nuclear power to a significantly higher level and initiates Thorium utilisation on a large scale.

There are a number of other aspects that also need attention. These are well covered in the report. The report is a valuable contribution to nuclear energy discourse in India which at times appears to have been hijacked by sectarian politics losing focus on core Indian interests. On issues like the civil nuclear liability, we seem to have actually created serious difficulties for Indian industries that had over years built globally competitive capabilities in nuclear technology in spite of tremendous odds. At a time when they should have started aspiring for making entry into the global market, they may well be having second thoughts on continuing with business in the nuclear area.

People at large as well as those in the neighbourhood of a major developmental activity, industry as well as key opinion makers are the prime stakeholders in any development context. They must be provided with credible inputs to keep the national consensus on the right track. I do hope that the comprehensive perspective in the nuclear energy presented in this report would be a valuable contribution in this context.

Anil Kakodkar
Former Chairman,
Atomic Energy Commission of India

Acknowledgements

This study would not have been possible without the generous support of the BRNS, DAE. The Indian Pugwash Society would like to extend its sincere gratitude to Prof Sangeeta for extending all the help and cooperation at all the stages of this study. The Society would like to acknowledge the support of Dr R.K. Sinha, Chairman DAE and Dr R.B. Grover, Director HBNI.

During the completion of this study, the research team of the society received valuable inputs from several DAE experts. The society would like to thank Dr K.L. Ram Kumar, Dr D.N. Sharma, Dr Pradeep Kumar, Mr Baburajan and Dr N. Nagaish. The society would also like to thank Dr Baldev Raj, Dr Chellapandi, Dr S.C. Chetal, and Dr Saibaba, for being generous in explaining technical and operational aspects of nuclear power installations in India.

Thanks are also due to Dr S.S. Bajaj, Chairman AERB and Dr Doraiswamy, Dy. Chairman AERB for interacting with society's research team on various dimensions of India's nuclear regulatory mechanisms.

The society is grateful to the NDMA chairman, Dr Shashidahr Reddy and Dr M.C. Abani for providing valuable insights into India's nuclear emergency preparedness. The society would also like to thank Prof B.M. Suri for his critical comments on the project manuscript and to Dr Anil Kakodkar for his guidance and encouragement to the activities of the society. The Society owes a debt of gratitude to Director General, IDSA for providing invaluable support to the Indian Pugwash Society in all the endeavours.

Abbreviations

ACPSR	Advisory Committees on Project Safety Review
ADS	Accelerator Driven Systems
AEC	Atomic Energy Commission
AERB	Atomic Energy Regulatory Board
AGSS	Aerial Gamma Spectrometry System
ALARA	As Low As Reasonably Achievable
ANSI	American National Standards Institute
ARS	Acute Radiation Syndrome
ASCE	American Society of Civil Engineers
ASE	Atomstroyexport
ASEAN	Association of South East Asian Nations
ASME	American Society of Mechanical Engineers
ASN	The French Nuclear Safety Authority
BARC	Bhabha Atomic Research Centre
BCM	Billion Cubic Meters
BDBA	Beyond Design Based Accidents
BDBE	Beyond Design Basis Events
BEIR	Biological Effects of Ionizing Radiations
BHNS	Bombay Natural History Society
BWR	Boiling Water Reactor
CAGR	Compound Annual Growth Rate
CCS	Carbon Capture and Storage
CERC	Concerned Electricity Regulatory Commission
CMG	Crisis Management Group
CNG	Compressed Natural Gas

CNS	Convention on Nuclear Safety
CO_2	Carbon Di-Oxide
CPP	Captive Power Plants
CSC	Convention on Supplementary Compensation
CSP	Concentrated Solar Power
CSR	Corporate Social Responsibility
CTBT	Comprehensive Testy Ban Treaty
DAE	Department of Atomic Energy
DBE	Design Basis Earthquake
DBGM	Design Basis Ground Motion
DDMA	District Disaster Management Authority
DIL	Derived Intervention Levels
DNA	Deoxyribonucleic Acid
DNI	Direct Normal Irradiation
DoE	Department of Environment
DRDs	Diamond Radiation Detectors
EAC	Expert Appraisal Committee
ECCS	Emergency Core Cooling System
EIA	Environment Impact Assessment
EIAN	Environment Impact Assessment Notification
EMS	Environmental Management System
ENEC	Emirates Nuclear Energy Corporation
EPR	European Pressurised Reactors
EPZ	Emergency Planning Zone
ERC	Emergency Response Centre
ERT	Emergency Response Teams
ESL	Environment Survey Laboratory
ESP	Environment Stewardship Programme
EU	European Union
FANR	Federal Authority for Nuclear Regulation
FBRs	Fast Breeder Reactors
FBTR	Fast Breeder Test Reactor
FDI	Foreign Direct Investment
FY	Financial Year

GDP	Gross Domestic Product
GHGs	Green House Gases
GIS	Geographical Information System
GWe	Giga Watt electrical
HBRA	High Background Radiation Areas
HEPA	High-Efficiency Particulate Air
HLLW	High Level Liquid Waste
HWB	Heavy Water Board
IAEA	International Atomic Energy Agency
ICJ	International Court of Justice
ICRP	International commission on Radiation Protection
IEEE	Institution of Electrical and Electronic Engineers
IEGC	Indian Electricity Grid Code
IERMON	Indian Environment Radiation monitoring Network
IGCC	Integrated Gasification Combined Cycle
IL	Intervention Levels
INES	International Nuclear and Radiological Event Scale
IPPAS	International Physical Protection Advisory Services
IRODOS	Indian Real Time Online Decision Support System
IWTMA	Indian Wind Turbine Manufacturers Association
JNNSM	Jawaharlal Nehru National Solar Mission
KAPS	Kakrapar Atomic Power Station
KGS	Kaiga Generating Station
KKNPP	Kudankulam Nuclear Power Plant
KWh	Kilo-Watt Hour
L&T	Larsen & Toubro
LNG	Liquefied Natural Gas
LNT	Linear No Threshold
LPG	Liquefied Petroleum Gas
MAPS	Madras Atomic Power Station
MEA	Ministry of External Affairs
MINT	Malaysian Institute for Nuclear Technology Research
MNPC	Malaysia Nuclear Power Corporation
MNRE	Ministry of New and Renewable Energy

MoEF	Ministry of Environment and Forests
MoH&FW	Ministry of Health and Family Welfare
MoU	Memorandum of Understanding
MRI	Mean Recurrence Interval
MRTS	Mass Rapid Transport System
MSM	Master Slave Manipulator
mSv	Mili Sieverts
MT	Mega Tonnage
MWe	Mega Watt electrical
MWeq	Mega Watt Equivalent
MWh	Mega Watt Hour
NAPCC	National Action Plan for Climate Change
NAPS	Narora Atomic Power Station
NCMC	National Crisis Management Committee
NDMA	National Disaster Management Authority
NDRF	National Disaster Response Force
NEERI	National Environmental Engineering and Research Institute
NGO	Non-Governmental Organisation
NNEA	National Nuclear Energy Agency
NO_2	Nitrous Oxides
NPCIL	Nuclear Power Corporation of India Limited
NPDSC	Nuclear Power Development Steering Committee
NPP	Nuclear Power Plants
NRC	Nuclear Regulatory Commission
NRPB	National Radiological Protection Board
NSDF	Near Surface Disposal Facilities
NSG	Nucler Suppliers Group
NSRA	Nuclear Safety Regulatory Authority
OBE	Operating Basis Earthquake
OECD	Organisation for Economic Co-operation and Development
OED	Off-Site Emergency Director
OEEC	Organisation for European Economic Co-operation
ONGC	Oil and Natural Gas Corporation
OPEC	Organisation of the Petroleum Exporting Countries

OSRAT	Operational Safety Review Team
PERCOSIB	Perchloric Acid Treated Siliceous Brick
PFBR	Prototype Fast Breeder Reactor
PHWR	Pressurized Heavy Water Reactor
PLF	Plant Load Factor
PMNAE	People's Movement Against Nuclear Energy
PSA	Probabilistic Safety Analysis
PSR	Periodic Safety Review
PSU	Public Sector Unit
PV	Photovoltaic
QA	Qualitative Assurance
R&D	Research & Development
RAPS	Rajasthan Atomic Power Station
RE	Renewable Energy
REC	Renewable Energy Certificates
RFPR	Resorcinol Formaldehyde Polycondensate Resin
RPO	Renewable Purchase Obligation
RPSP	Renewable Purchase Specification
RSW	Radiation Shielding Window
SAR	Safety Analysis Report
SARCOP	Safety Review Committee for Operating Plants
SBO's	Station Black-Out Scenario
SDMA	State Disaster management Authority
SDR	Special Drawing Rights
SED	Site Emergency Director
SER	Site Evaluation Report
SERC	State Electricity Regulatory Commission
SHP	Small Hydro Power
SORC	Station Operation Review Committee
SPCB	State Pollution Control Board
SSE	Safe Shutdown Earthquake
TAPI	Turkmenistan, Afghanistan, Pakistan and India
TAPS	Tarapur Atomic Power Station
TCO	Techno-Commercial Offer

TERI	The Energy Research Institute
TES	Thermal Ecological Studies
VARNSAC	The Vietnamese Agency for Radiation and Nuclear Safety and Control
TLDs	Thermo Luminescent Dosimeters
TMC	Tata Memorial Cancer Centre
TMI	Three Mile Island
TPCES	Total Primary Commercial Energy Supply
UAE	United Arab Emirates
UCG	Under Ground Coal Gasification
UK	United Kingdom
UNSCEAR	United Nations Scientific Committee on the Effects of Atomic Radiation
US	United States
VVER	The Voda Voda Energo Reactor
VWP	Vitrified Waste Product
WANO	World Association of Nuclear Operators
WIL	Walchandnagar Industries Limited
WIP	Waste Immobilisation Plant
WISE	World Institute for Sustainable Energy
WNA	World Nuclear Association
μSv	microsieverts

About the Editors

Dr Arvind Gupta currently is the Director General of the Institute for Defence Studies and Analyses (IDSA) and Convenor of the Indian Pugwash Society. Earlier he held the Lal Bahadur Shastri Chair, at the IDSA. He holds a PhD in International Relations from the School of International Studies, Jawaharlal Nehru University, New Delhi. A member of the Indian Foreign Service, Dr Gupta served in Indian Missions in Moscow, London and Ankara. He was Joint Secretary in the National Security Council Secretariat (NSCS) from1999 to 2007. He has number of authored and edited books to his credit including the IDSA-Indian Pugwash Society Publication, *India in a Changing Global Nuclear Order* in 2009 and *Emerging Asian Nuclear Environment: Implications for India (2012).*

Dr. K. D. Kapur is a former Professor of Diplomacy, School of International Studies, Jawaharlal Lal Nehru University, New Delhi. Currently, Prof. Kapur is Indian Council Social Science's (ICSSR's) Senior Fellow, at the Institute for Defence Studies and Analyses (IDSA). Prior to this Prof. Kapur was associated with the Indian Pugwash Society as a Consultant. He was a Visiting Professor (2007-2010) at the Academy of International Studies, Jamia Milia Unmiversity, New Delhi. He was a Senior Fellow (1976-1977) at the Kiev State University, (former Soviet Union) and a Japan Foundation Fellow (1995-1996) at the Reitaku University, Chiba Ken, Japan. He was associated with *IDSA* during 1984-85 as ICSSR Senior Fellow. He has authored four books including *Nuclear Non-Proliferation Diplomacy – Nuclear Power Programmes in the Third World* (1993); *Soviet Nuclear Non-Proliferation Diplomacy and the Third World* (1993); *Nuclear Diplomacy in East Asia* (1995) and co-edited a volume with Dr Arvind Gupta, titled, "*Emerging*

Asian Nuclear Environment: Implications for India (2012). He has carried out a number of UGC/ICSSR sponsored research projects including, *Political Economy of Multilateral Trade Negotiations* including interface between Trade and Environment; *India's Regional Trading Arrangements in East Asia; and High-Tech Transfers to India.*

G. Balachandran is a Consulting Fellow at the IDSA. He has a PhD in Economics and Computer Science from the University of Wisconsin. His specialization includes issues related to economics and technology. He is also associated with the National Maritime Foundation, New Delhi. He is a Consultant both to the Americas Division of the Indian Ministry of External Affairs and to the Confederation of Indian Industry.

Kapil Patil is a Research Associate at the Indian Pugwash Society. He holds an MPhil in Disarmament & Diplomacy Studies from Jawaharlal Nehru University, New Delhi and Masters Degree in Strategic Studies from University of Pune. He has contributed research articles on India's nuclear energy programme, nuclear safety policy, civilian nuclear liability etc. He has participated in Nuclear Proliferation International History Project (NPIHP) Boot Camp in May 2011.

List of Tables and Figures

Tables

Figures

Introduction

Arvind Gupta

India's nuclear energy programme is at a crossroads. The Fukushima nuclear accident led the government to review the safety features of the Indian nuclear reactors. The safety review unambiguously concluded that India's nuclear power plants are safe and wield sufficient safety margins to deal with extreme natural events as witnessed in Fukushima, Japan in 2011. The Atomic Energy Regulatory Board (AERB) conducted the safety review, which though a professional body, still functions under the aegis of Department of Atomic Energy (DAE). For this reason, the public confidence in the AERB's safety review might be low. However, the Indian public and the policymakers have traditionally been supportive of the country's nuclear energy programme. Over the years, successive governments have invested hugely in country's nuclear energy infrastructure. As a result, India has developed robust scientific and industrial capabilities in the nuclear energy sector.

Yet, post Fukushima, the public confidence in India's nuclear programme has been shaken. This is also a part of the world-wide wave of scepticism regarding nuclear energy. India is no exception. Public protests against nuclear energy have been witnessed in different parts of India. While some protests may have been organised by vested interests, others are motivated by genuine safety concerns. The protests, genuine or not, have given rise to a public debate on the desirability of nuclear energy in India. The protagonists of nuclear power offer an array of arguments. India, being an energy deficient country requires nuclear power for growth and development; nuclear energy has been cost competitive vis-à-vis the alternatives; and it is reasonably safe with minimal carbon footprints.

The opponents of nuclear energy, on the other hand, raise serious concerns over India's nuclear energy programme. They primarily argue that nuclear energy is hazardous, citing concerns over safety of nuclear reactors, radiological waste disposal, and challenges involved in maintaining adequate preparedness to deal with nuclear reactor accidents. They have been critical about the lack of transparency in India's nuclear energy programme, bemoaning that the public has little access to relevant information. Most importantly, the critics have strongly pitched for autonomous functioning of the regulatory agency, the Atomic Energy Regulatory Board (AERB). India's nuclear energy programme has failed to deliver what was promised by the scientists; as nuclear energy constitutes a miniscule 2.5 per cent of the total electricity production in the country which is much less than that produced by renewable energy sources. The opponents are therefore, sceptical about the future of nuclear power generation in the light of the delays and procrastinations of the past.

This report is being written at a time when the future of nuclear energy globally, looks less than bright. The nuclear renaissance that the IAEA talked about some years ago appears to be petering out. This report takes the view that nuclear energy programme in India must continue. Nuclear energy must remain a part of India's energy mix. The safety management of nuclear installations in India has been competent over the years. The safety record of India's nuclear power programme has been reasonably satisfactory. The three major nuclear accidents—the Three Mile Island, Chernobyl and Fukushima—were mostly caused by human negligence. The AERB should therefore derive appropriate lessons from these events and implement safety measures more rigorously. The use of layered safety systems in modern day nuclear reactors has significantly minimized the safety risks. The waste disposal issues are being handled by adopting international best practices. This has resulted in significant reduction of risks associated with nuclear energy to people and the environment.

This study suggests that public concerns relating to nuclear energy need to be taken seriously and addressed properly through far greater transparency, accountability and dialogue. The criticism that the nuclear energy establishment has failed to fully engage with the public is valid. Credible and appropriate information needs to be put out in the public domain. The nuclear regulator should be made independent. There should be a periodic safety audit of the various programmes by the independent regulator and the public should be given a correct assessment of the strengths and weaknesses of country's nuclear programme. Can India realistically achieve the nuclear energy generation targets set out for 2050, given that there have been major slippages in the past, and that the current international environment is hostile to nuclear energy? What should be the realistic target? The current as well as the future energy targets must be reviewed in order to make them realistic and achievable.

While the political parties and the public at large have supported country's nuclear programme in the past, there is a need for greater transparency and accountability in nuclear energy management. The study analyses range of public concern regarding various aspects of nuclear energy which have assumed immense social and political significance in India in recent years. It shows that the public fears surrounding the safety of nuclear reactors appear somewhat exaggerated, especially after the Fukushima accident. Technical experts do not agree with the inflated risks perceptions of common people and the environmental groups. There is overwhelming scientific evidence which indicate that risk perception among the lay people has no particular relation to the actual hazard ranking of nuclear energy. The study provides an assessment of the overall performance of India's nuclear power plants vis-à-vis the concerns expressed by people and the civil-society groups.

The study evaluates India's overall energy requirements and highlights the role that nuclear energy can play vis-a-vis other sources of energy. It examines the country's nuclear power programme on various parameters such as reactor safety, environmental and radiation protection measures, nuclear waste management practices, nuclear emergency preparedness and issues pertaining to civilian nuclear liability. The assessment is carried out based on the extensive interviews and discussions held with various DAE scientists, NPCIL engineers, and the AERB officials in Mumbai. Besides, numerous scientific publications were consulted to assess the compliance of Indian reactors with the AERB guidelines and regulations. The field visit to Tarapur Atomic Power Station (TAPS) and its Environment Survey Laboratory (ESL) has provided useful insights into the radiation survey monitoring procedures and environmental protection methods employed at the NPPs. The evidence gathered along with the existing public knowledge on these issues has enabled a greater understanding on various aspects of India's nuclear energy programme.

The first chapter of the book, 'Post Fukushima Global Nuclear Scenario', evaluates the impact of Fukushima accident on the quantum of nuclear energy produced worldwide and its share in the global energy mix. It reviews the civil nuclear energy programmes in different countries, including those countries which have decided to abandon their nuclear energy programmes. The chapter underscores that with the exception of certain European countries, nuclear power programmes all over the world continues uninterrupted, despite the Fukushima crisis. More than 60 nuclear reactors are currently under construction in thirteen different countries across the globe. However, post Fukushima, the safety of the nuclear reactors has emerged as a prime concern forcing many countries to review the safety of nuclear power plants.

The second chapter, 'India's Energy Mix: Energy Security and Economic Development', deals comprehensively with the interface between energy security and economic development. It focuses on total electricity generation in India from fossil fuels (coal, oil and gas); from non-fossil fuels (wind, hydro, solar, biomass and nuclear); and their environmental impacts. It then analyses the contribution that nuclear energy can make towards ensuring India's energy security, meeting the challenges of climate change and fostering economic growth.

The third chapter, 'Public Concerns over Nuclear Power in India', highlights the specific concerns of the common people with regard to new nuclear projects in different parts of India. It examines the influence of concerns such as nuclear safety, radioactive waste management, perceived knowledge about the consequences of nuclear energy, etc. It broadly divides public concerns into three major categories: (i) safety and radiation related concerns; (ii) environmental concerns; and (iii) socio-economic concerns. The chapter makes a detailed assessment of various risk management approaches adopted by the Department of Atomic Energy to assuage concerns and enhance public acceptance of nuclear power in India. It contends that socio-economic concerns prominently shape lay people's attitude or beliefs about nuclear energy and importance of addressing such concerns while harnessing nuclear power.

The fourth chapter, 'Safety of India's Nuclear Power Plants', reviews the nuclear reactor safety management processes in Indian NPPs as mandated by the AERB. It discusses in detail various safety measures incorporated in Indian NPPs after the Fukushima nuclear crisis. The chapter argues that the lessons learned from the Fukushima nuclear accident highlight the need for improving safety performance of systems, structures and components in NPPs to withstand severe external events. It underscores that the NPPs in India are designed using the 'defence-in-depth' approach to cope with diverse challenges presented by extreme natural events.

The fifth chapter, 'Environmental Impact of Nuclear Power Plants in India', discusses the environmental risks associated with nuclear energy. It analyses how the risks are addressed in various stages of nuclear power reactor design, construction and operations. It reviews the regulatory requirements for siting NPPs and describes the evolution of legal framework for environmental impact assessment for nuclear installations in India. It highlights that the process of granting environmental clearances for NPPs has considerably evolved over the years and is on par with international best practices. It also shows that the operation of NPPs in India have not caused any discernible hazard to the environment as well as the surrounding population.

The sixth chapter, 'Nuclear Waste Management Strategy of India', primarily focuses on various challenges of nuclear waste management. It discusses India's nuclear waste management strategy which is primarily aimed at minimising waste generation at all stages of design, operation and maintenance of the NPPs. It contends that waste volume reduction is a key factor for the effective confinement of waste by using various processes and matrices for the treatment and conditioning of radioactive waste. The chapter reveals that India has rich experience comparable with the best international standards for management of radioactive waste from nuclear power plants, fuel reprocessing and allied installations. The current waste management practices have reasonably achieved the prime objective of ensuring the safety of the environment through suitable treatment and conditioning of the nuclear waste.

The seventh chapter, 'Radiation Releases from Nuclear Power Plants in India', focuses on radiation releases from NPPs and fuel cycle facilities in India. The chapter describes the effects of low level and high level radiation on public health as established by various studies conducted by the International Council of Radiation Protection (ICRP) and the United Nations Scientific Committee on Effects of Atomic Radiation (UNSCEAR). It reviews the impact on human health caused by major as well as minor accidents at nuclear and radiological facilities all over the world. Also, it evaluates the radioactive releases from Indian NPPs and reviews the epidemiological studies as conducted on radiation and non-radiation workers and their families. The chapter concludes that radiation doses received by the nuclear plant workers are significantly lower than the prescribed regulatory limits and pose no serious health risks to them and their families.

The eighth chapter, 'Nuclear Emergency Preparedness in India', deals with emergency preparedness plans and capabilities developed by the Department of Atomic Energy (DAE) and the National Disaster Management Authority (NDMA) to effectively manage and mitigate the off-site nuclear emergency. The chapter focuses on the lessons of the Fukushima crisis to improve off-site emergency preparedness in the form of advance technological systems, new organisational structures and the development of infrastructure in areas around the NPP's. It also identifies areas where India needs to augment capabilities to deal with off-site emergency at the nuclear power installations.

The ninth chapter, 'Civil Nuclear Liability in India: Challenges and Prospects', discusses in detail, the salient features of the international nuclear liability regime. It analyses the prospects and challenges for India's nuclear industry in view of the Civil Liability for Nuclear Damage Act of 2010. The chapter discusses the implications of this Act for international as well as domestic suppliers to India's nuclear energy programme. It suggests certain structural adjustments and remedial measures to deal with the challenges presented by India's civilian nuclear liability act. The concluding chapter presents the main findings of the book and aims to stimulate meaningful discussion on the desirability of nuclear energy in India.

CHAPTER 1

Post-Fukushima Global Nuclear Scenario

K.D. Kapur

Introduction

The Fukushima nuclear accident has once again raised the question of the viability of nuclear energy as a safe energy option for the energy hungry countries of the world. Its credentials as a renewable source of energy in the overall energy mix of the countries have been well established. It is considered to be least polluting, and is environment friendly. It is now well recognised that during the next century there will a huge demand for energy to meet the needs of the growing population, and to lift billions of people out of poverty. Will nuclear energy industry survive the Fukushima nuclear accident? This is the single most important question facing governments and industry.

Nuclear energy over the years has contributed significantly towards reducing reliance on fossil fuels in many countries and has been considered an important component in the energy basket of the countries. Even nuclear waste is believed to cause very little environmental pollution. Whereas coal waste is dispersed, nuclear waste can be put into cask-storage and considerably reduced by adopting a close end fuel cycle. Electricity is needed for industrial, agricultural growth and for a good life. The increasing fossil fuel prices have made it all the more imperative that nuclear energy is accepted as a viable option. In view of the constant and gradual depletion of the non-renewable sources of energy, it is critical that the share of the renewable sources of energy, including nuclear energy, should be increased. In this respect, it can be argued that nuclear energy is one of the best available options. Though there is little likelihood of a nuclear renaissance, but the pursuit of nuclear energy in the developing countries is likely to continue in the near future, to ensure energy security but with greater emphasis on safety.

This chapter attempts to identify the status of nuclear power at the global level after the Fukushima Daiichi nuclear accident. It proposes to discuss the global nuclear scenario and the attitude of various countries towards nuclear energy after the Fukushima accident. It will also examine whether countries have turned their backs on the nuclear energy. The chapter has been divided into following sections: Introduction; Global concerns regarding the safety of nuclear power plants; Japan's attitude towards nuclear energy after the Fukushima nuclear accident; Is Fukushima a setback to the global nuclear renaissance?; Status of nuclear energy at the global level before the Fukushima and after the Fukushima nuclear accident; Response of various countries; and Conclusion.

Global Concerns about the Safety of Nuclear Power Plants

The nuclear disaster in Japan's Fukushima Daiichi nuclear power plant consequent to massive earthquake and tsunami seemed to have delivered a body blow to the prospects of the development of nuclear energy at the global level. Immediately after the crisis many countries that were planning to procure nuclear power plants started having second thoughts. Those who were already in the process of building new reactors stressed the need for reviewing the safety standards. Even the Director General IAEA Yukiya Amano sounded a note of pessimism by saying that the global nuclear community cannot take a "business as usual approach" and lessons must be learnt from Fukushima nuclear crisis as the plant has been leaking radiation into the environment ever since it was hit by a massive tsunami. Speaking to reporters he said, "I am confident that valuable lessons will be learned from the Fukushima Daiichi incident, which will result in substantial improvement in nuclear operating safety, regulation and the overall safety culture".[1] The issue of improving the safety of the nuclear reactors has assumed a great importance after the Fukushima nuclear accident. The Three Mile Island, Chernobyl, and now Fukushima nuclear accidents are a reminder that nuclear safety requires constant vigilance. The target of tripling the nuclear energy capacity by 2050 demands that risks of nuclear accidents be reduced considerably and the least safe facilities must become the focus of attention. The Fukushima nuclear accident has generated a debate within and outside Japan about the viability of nuclear energy as an option for sustainable economic development and the energy security of countries, worldwide. In many countries around the world and particularly in the European Union and the Scandinavian countries the focus is on how to increase the share of non-polluting renewable energy sources so as to ensure energy security.

The Fukushima nuclear crisis has also led to the reinvigoration of anti-nuclear movements around the world, mainly in the developed countries, and also in India. There is an upsurge of anti nuclear protests in India, particularly, against the six European Pressurised Reactors (EPR) that are being imported from France and are to be located in Jaitapur, Maharashtra. The commissioning of two reactors built with Russian assistance at Kudankulam too has been delayed because of agitations by the anti-nuclear NGOs. The anti-nuclear agitation is supposed to have been mainly engineered and supported by foreign NGOs through their local contacts. The opposition to the nuclear reactors is based on variety of reasons such as: safety of nuclear reactor design, nuclear radiation, environmental concerns, waste management and disposal, health hazards to the surrounding population, uprooting of the local population, the damage to the fisheries leading to unemployment, and land acquisition. However, the government of India has assured the people in these areas that all their concerns would be addressed and the safety of the nuclear power plants would be the highest priority of the government.

Japan's Attitude towards Nuclear Energy after the Fukushima Accident

The Japanese government's stance on nuclear energy has undergone a considerable change after the Fukushima nuclear accident. In May 2011, the Japanese prime minister, Naoto Kan had announced that he would scrap all the plans that stipulated the building of 14 new reactors by 2030 saying that the country needed to "start from scratch". In his view Japan needed a new energy policy. The plans also envisaged an increased share of nuclear power in Japan's electricity supply—from 30 to 50 per cent. The prime minister's statement indicated a major shift in Japan's nuclear policy whereby Japan's energy policy in the future would include two additional pillars i.e. renewable energy and conservation. He also emphasised the "need to make nuclear energy safer and do more to promote renewable energy". It also marked a change from his earlier stance when he had suggested that Japan should sell nuclear technology to emerging nations as a new source of export income.[2] In July 2011, Kan went a step further and said that Japan should "reduce and eventually eliminate dependence on nuclear energy as Fukushima had demonstrated dangers of nuclear technology".[3]

However, almost a year later Naoto Kan's successor Yoshihiko Noda departed from Kan's position and adopted a pro-nuclear stance on the role of nuclear energy. He painted a brighter future for the nuclear energy and wanted it to be an important component in the country's energy policy. Whereas Kan had called for ending Japan's dependence on nuclear power Noda pointed out that Japan needed nuclear power to prop up it's sliding economy by providing energy security. Earlier in December 2011 he had announced that technicians had taken control of the Fukushima Daiichi nuclear Power Plant.[4] By March 2012 all but two of Japan's 54 commercial reactors had gone off line for safety checks and the last reactor was shut down some time in April thereby temporarily shutting down an industry which generated almost one third of Japan's electricity.

Prime Minister Noda, however, has assured that the reactors will be started as soon as possible but with the approval of the local community leaders. He also supported a gradual phasing out of nuclear power over several decades. [5] The situation as of March 2013 remains the same—only two of Japan's 54 nuclear reactors are in operation. The rest are still closed for maintenance and repairs resulting in zero nuclear power supply. The present Prime Minister, Shinjo Abe, is aware of the importance of nuclear energy for the country's energy security, and is determined to reverse the decision of his predecessor and restart the reactors as soon as possible. However, he is emphasising the tightening of safety requirements. The government of Prime Minister Abe is expected to unveil a new energy policy by the end of 2013 in which country's dependence on nuclear energy is likely to be reduce considerably.

Despite the Fukushima nuclear accident and the mounting public criticism, the previous Prime Minister, Noda supported the export of nuclear reactors to Vietnam, Jordan and Lithuania. This has put Japan in a precarious position, as it is trying to reduce its reliance on nuclear power on the one hand while at the same time it is actively supporting the nuclear export efforts of its domestic nuclear industry. Japan's industrial heavyweights like Toshiba Corp, Hitachi Ltd and Mitsubishi Heavy Industries Ltd have succeeded in convincing the government that their technologies can still win contracts worth billion of dollars at a time when the Japanese economy is in doldrums. The Japanese industrial conglomerates have been trying to finalise deals by highlighting the safety standards of its nuclear reactors. Toshiba and Hitachi had a hand in the building of Fukushima plant reactors and both of them are supposed to have lost business in many countries. They pulled out of a joint bid in Turkey. However, they believe that the setback is minor as before the Fukushima earthquake, Hitachi had plans to construct 38 reactors by 2030 and although the achievement of this target, may be bit delayed, the company, has not formally lowered the target. Japanese companies are aware of the fact that the demand for atomic energy in many parts of the world has not decreased despite the decision of Germany to phase out all its nuclear reactors by 2022. The other significant development is Turkey's recent vote against construction of new nuclear reactors. But at the same time many countries in Asia and the Middle East still have robust plans for major nuclear projects.[6] Hitachi officials do not think, "that Fukushima poses a threat to their negotiations" and believe that "in terms of energy security and mitigating carbon emissions, there is a still major demand."[7]

Japan's Nuclear Exports and Bilateral Nuclear Cooperation

In a bid to convince its customers of the safety of its reactors Hitachi sent its executives to various countries to reassure the governments regarding pending bids and deals. It particularly focused on Lithuania where it had made a joint bid with the US based General Electric Co. to build a 1,300 MW facility. Lithuania closed its last Soviet designed reactor in 2009. In June 2011 Hitachi officials offered Lithuania new safety measures devised since the Fukushima accident that included changes in the backup power system. Finally, in July 2011 Hitachi signed an agreement with Lithuania for the construction of a nuclear reactor.[8]

In a major development Prime Minister Noda succeeded in signing bilateral civilian nuclear cooperation accords with Jordan, Russia, South Korea and Vietnam that were approved by the Diet on December 9, 2011 after the opposition controlled Upper House passed the bill. These agreements were signed before the Fukushima nuclear accident of March 11, 2011. This paved the way for Japan to export its nuclear technology to the four countries. Japanese companies hope to export nuclear power plants to Jordan and Vietnam and reactor parts to South Korea. A nuclear cooperation agreement with Russia could give Japan access to uninterrupted nuclear fuel supply.[9] Interestingly, the Japanese foreign minister Koichiro Genba told the foreign affairs panel of the House of Representatives on December 1, 2011, that Japan was also hoping to conclude bilateral accords on cooperation for the peaceful use of nuclear energy with five other countries—Brazil, India, South Africa, Turkey and the UAE and talks with these countries began after the Fukushima nuclear accident.[10]

A month after shutting down all the 50 nuclear reactors for safety review, Prime Minister Noda's government on June 17, 2012 decided to restart reactors, 3 and 4, at the Kansai Electric Power Company's Ohi nuclear plant. The utility serves the one trillion dollar economy of Japan's second biggest urban region. Other plants too are likely to be restarted. The mayor of Ohi and the governor of Fukui, where other nuclear power plants are located, also supported the government's decision and wanted that reactors at their nuclear power plants too should be reactivated soon.[11] The 1,189 MW Reactor 3 started generating power from July 5, 2012, four days after it was started.[12]

Despite the overwhelming opposition[13] to nuclear energy and the setting up of nuclear plants, the government decided to go ahead and restart the nuclear reactors because of concerns relating to the mounting demand for power as the summer months got underway. The government, however, insisted that its number one concern was the safety of the reactors. On June 8, 2012, Prime Minister Noda made a personal appeal to the nation to agree to the resumption of the Reactors 3 and 4 at Ohi arguing that:...in order to lead prosperous and decent lives, cheap and stable electricity is indispensable. Japanese society will not be able to function if there is a decision to permanently halt nuclear power generation, which has accounted for approximately 30% of our total electricity supply or if nuclear power generation remains halted.[14]

Is Fukushima a Set Back to the Global Nuclear Renaissance?

The nuclear renaissance, if at all there was one, has been confined only to the developing countries, as nuclear energy in OECD countries has been either stagnant or decreasing over the last couple of years and virtually no new nuclear reactors have been built in these countries. As far as the non-OECD countries are concerned, despite the caution and desire to learn from the Fukushima nuclear crisis in terms of safety and nuclear reactor design, most countries have declared their commitment to nuclear energy. Thus there has been little impact on the nuclear renaissance, as most of the developing countries are still continuing with their nuclear power programmes.

It is however, certain that these countries will have to face opposition from environmentalists and other civil society organisations. Even among non-OECD countries the bulk of the nuclear reactors are to be constructed in two countries, China and India as they have ambitious nuclear power programmes. Both have expressed their commitment to nuclear energy in the wake of rising petroleum prices. In addition to India, China, and South Korea a number of new aspirants for nuclear energy have emerged in Asia. These include Vietnam, Myanmar, Indonesia, Taiwan, UAE, Saudi Arabia, Egypt, Iran, and Bangladesh. It is pertinent to mention here that despite the Fukushima nuclear accident and the pessimism of some European countries and environmentalist groups with regard to nuclear power, the protagonists of nuclear energy are quite optimistic about the future of the nuclear energy.

Nuclear Energy at the Global Level

Even at the global level very high targets are being set for nuclear energy. After the shutting down of six nuclear reactors at the Fukushima Daiichi nuclear power plant, on March 1, 2012, the total number of nuclear power reactors under operation worldwide has come down to 436 from 442, with a total net installed capacity of 370.128 GWe. These reactors are operating in 31 countries around the world. Five reactors are in long term shut down and 63 nuclear reactors are under construction.[15] These 63 have an installed capacity of 61 GWe and are spread over 15 countries. Of these 26 are in mainland China, two in Taiwan, ten in the Russian Federation, five in the Republic of Korea, six in India, two each in Japan, the Slovakian Republic, Ukraine and Bulgaria and one each in Argentina, Brazil, Iran, Finland, France, Pakistan and the United States.[16]

According to World Nuclear Association (WNA) as of December 2011, nuclear power provided about 13.5 per cent of the world electricity; almost 24 per cent of the electricity in the OECD countries; and 34 per cent in the EU.[17] Nuclear generation is an important component of the global electricity mix with coal contributing 40 per cent; oil 5 per cent; natural gas 21 per cent; and hydro and others 20 per cent. Nuclear electricity is suitable for large scale, continuous electricity demand which requires reliability (i.e. base load)[18] Out of these 31 countries about 16 are dependent on nuclear power for at least a quarter of their electricity with France getting about 74.1 percent of its total electricity from the nuclear power. In most of the developed countries the nuclear share in the electricity generation constitutes a sizable percentage. The table below will vouch for it.

Table 1.1: Percentage of Nuclear Power Generation in Various Countries

Armenia	39.4	Belgium	51.1
Bulgaria	33.1	Canada	15.1
Czech Republic	33.3	Finland	28.4
Germany	28.4	Hungary	42.1
Japan	29.2	Russia	17.1
South Korea	32.2	Romania	19.5
Slovakia	51.5	Slovenia	37.3
Spain	20.1	Sweden	38.1
South Africa	5.2	Taiwan	19.3
Switzerland	38.0	Ukraine	48.1
USA	19.6	UK	15.7

The share of nuclear energy in the developing countries' energy mix is quite low and only a few countries have around a three percent share. For instance India's share is a meagre 2.9 percent; China 1.8 percent; Mexico 3.6 percent; Pakistan 2.6 percent; and Argentina 5.9 percent.[19] Several new entrants have begun construction of nuclear reactors while several other states are keen to go nuclear. In addition 56 countries operate 250 research reactors and 220 nuclear powered ships and submarines. It is also a fact that most of the new nuclear reactors are now being built in India and China with China having an almost 60 per cent share. The developed world, having strongly reacted to the Japanese nuclear disaster, is likely to freeze any future nuclear power development. In view of this, the Chinese and Indian nuclear markets are likely to become even more important for companies such as Westinghouse, which is designing most of the Chinese reactors, and Areva and others, who are designing for India. Both countries will emerge as the hot destinations for foreign companies to market nuclear reactors.[20]

Following the Fukushima nuclear accident the *World Energy Outlook 2011* scaled down the increase in nuclear capacity from 90 per cent as stipulated in the *World Energy Outlook 2010* to 60 per cent by 2035. Global nuclear power projections as per New Policies Scenario are expected to touch 630 GWe by 2035 around 20 GWe lower than the 2010 projections. Despite this, nuclear power projections continue to play an important role, for providing base load electricity. It is indicated that most non-OECD countries and even many OECD countries are expected to go ahead with plans to install additional nuclear power plants even though there may be short term delays as the safety standards of the existing plants are reviewed.

The World Energy Outlook, 2011 also includes the Low Nuclear Case scenario where the role of nuclear energy gets scaled down, which in turn would have implications for the global nuclear energy balance. Assuming that the nuclear energy generation comes down to 335 GWe by 2035 the share of nuclear energy will drop from 13.5 percent to 7 percent by 2035 in the overall global energy mix. This could have serious implications for, "energy security, diversity in fuel mix, spending on energy imports and energy related CO_2 emissions". Its effect would be to 'increase import bills, heighten energy security concerns and make it harder and more expensive to combat climate change".[21]

Interestingly, the *Energy Roadmap 2050*, published by the European Commission in December 2011, revealed a positive attitude towards the nuclear power and said that nuclear energy can make "significant contribution to energy transformation process", and is "a key source of low carbon electricity generation that will keep system costs and electricity prices lower". As a large scale low-carbon option, nuclear energy will remain in the EU power generation mix". The 2050 energy roadmap examined the possible scenarios which would help EU achieve low carbon economy by 2050. It projected an 80 per cent reduction in the CO_2 emissions, which could be achieved by energy efficiency, renewable, nuclear power, and carbon capture and storage (CCS). In all this, electricity will play a major role and would almost double its share in final energy demand between 36 and 39 per cent in 2050.[22]

Earlier IAEA, in its report in 2010 had significantly raised its projections of the world's nuclear generating capacity. It projected that by 2020 a net 73 GWe new capacity would be added worldwide. The IAEA projection gave nuclear power a 13.5 per cent to 14.6 per cent share in electricity production in 2020 and 14.6 per cent to 15.9 per cent in 2030 with the fastest growth in Asia. However, this now is likely to be scaled back to some extent following the Fukushima accident, notably in Europe and Japan. Elsewhere there would be no change but some delay is likely due to reappraisal of regulations. At the global level, if such high targets are to be achieved, nuclear energy must become attractive for utilities, governments, and the public around the world. It should be seen as being: cost-efficient, less accident prone, safe against terrorist sabotage, less polluting and immune to nuclear proliferation. "In short, safety, security, non-proliferation and waste management are essential enablers for large scale nuclear energy growth."[23] More Chernobyl and Fukushima type catastrophes or a terrorist attack would have the potential to doom the prospects of nuclear growth on the scale needed to deal with climate change.

What is needed in the long term are new reactors and fuel cycle designs that are "cheaper, safe, more easily secured, more proliferation resistant, and more appropriate for developing counties with modest electricity grids and technical infrastructures could have major impact on nuclear energy's role in the carbon mitigation".[24] The issue of improving safety of the nuclear reactors has assumed great importance. The Three Mile, Chernobyl, and now Fukushima nuclear accident are reminders that nuclear safety requires constant vigilance.

In the first instance, the aging first generation nuclear reactors design pose a serious safety risk as they still lack containment vessels and emergency core cooling systems. So their licences should be renewed only after ensuring that necessary steps have been taken to ensure that these reactors do not pose any serious risk of radiation release. Second, there is need to help the new entrants as they do not have experience of operating effective nuclear regulatory systems, create a sound nuclear safety culture and have well trained and capable personnel. One way could be to supply these countries with small factory built reactors with high levels of built-in safety and security which could be deployed at a site and generate electricity for 10-20 years.

India and China with their ambitious nuclear power expansion plans have to pay special attention to these aspects. Global efforts were initiated after the Chernobyl nuclear accident to ensure the safety of nuclear reactors. This has resulted in the emergence of a broad international nuclear safety regime. This regime includes the Convention on Nuclear Safety, variety of agreements on nuclear liability in the event of nuclear accidents, a set of non binding norms and international standards and a web of organisations to promote it. The IAEA has developed a series of safety standards and guidelines for the safety of nuclear reactors which although do not carry the force of international law but are widely accepted and followed. The IAEA also offers peer reviews safety arrangements at individual reactors. The World Association of Nuclear Operators (WANO) consisting of nuclear power reactor operators from all over the world exchange information on nuclear incidents, the lessons learned and best safety practices. The IAEA and OECD's Nuclear Energy Agency manages a global safety incident reporting system. In addition, there are bilateral and multilateral nuclear safety assistance programmes, international professional associations and other groups focused on nuclear safety.[25]

Despite the international safety regime being quite broad based there are still some gaps in it. The Convention on Nuclear Safety does not have any binding standards for the safety of nuclear facilities. The IAEA peer review takes place only when a state asks for it; and in practice most of the nuclear reactor operators have never asked for it. Therefore, the IAEA would have no answer if asked which nuclear reactors are accidents prone. It has no real way of knowing which reactors in the world pose a serious risk. The WANO reviews are more or less universal but are not rigorous enough and operators usually keep them confidential and do not always implement all the recommendations. Both IAEA and WANO have warned that safety incidents are not being reported. In 2008 the IAEA's Commission of Eminent Persons in its report titled, 'Reinforcing Global Nuclear Order, suggested the following measures: (a) IAEA should lead efforts to establish "a global nuclear safety network" that would strengthen exchange of safety critical knowledge, experience, and lessons learned; (b) Over time "states should enter into binding agreements to adhere to effective global safety and standards and to be subject to international nuclear safety peer review"; (c) the IAEA and relevant states should greatly strengthen their efforts to help newcomer states "develop sound safety infrastructure"; and (d) the IAEA should expand its efforts to help states around the world assess and strengthen nuclear safety culture.[26]

Country Specific Response to the Fukushima Nuclear Accident

The response of different countries following the Fukushima nuclear disaster is as follows:

China: The focus countries of the nuclear renaissance—China and India—reacted cautiously to the Fukushima nuclear accident. China has an ambitious nuclear power programme with 14 nuclear reactors in operation and other 31 under construction.[27] It has 10.8 GWe installed capacity and a target of

70 GWe by the year 2020. Though China has declared that they will not to change their plans for building nuclear power units, it has nonetheless committed itself to stringent safety measures for its nuclear reactors apart from updating current nuclear safety regulations and guidelines. The nuclear projects not in compliance with nuclear safety guidelines are likely to be either suspended or terminated. Immediately after the Fukushima accident it declared a moratorium on the construction of nuclear reactors and asked for an intensive review of safety. At the same time, it also reiterated its commitment to nuclear energy.[28] The Chinese vice minister of environmental protection, Zhang Lijun, said that though China will learn from Japan's experience, "it will not change its determination and plans for developing nuclear power".[29] The Chinese believed there was a positive dimension to the Fukushima nuclear crisis as it would encourage the "healthy development| of the Chinese nuclear industry in the long run as it would be compelled to take additional safety measures" according to Cao Peixi Chairman of Huaneng Power International Inc.[30] In March 2012, Xinhua reported that: (i) China has ordered the safety review of all its nuclear power plants as it was conscious of the "inevitability" of the risks involved; [31] (ii) it was set to resume the construction of all power plants and (iii) and the importance of the nuclear energy for the energy security of the country.[32]

India: The Fukushima nuclear accident has once again shifted the focus regarding nuclear safety to the popular level. The Japanese nuclear crisis has raised safety concerns about India's nuclear reactors. The Indian government, committed as it is to nuclear energy, in a quick response assured the nation that all safety aspects of the reactors would be addressed and the safety would be updated in view of the lessons learnt from the Japanese experience. The Atomic Energy Commission (AEC) and Nuclear Power Corporation of India Limited (NPCIL) assured the nation that all nuclear power plants are absolutely safe. The Indian prime minister also assured the Parliament and promised an "immediate technical review of all safety systems" of all Indian power plants particularly in the context of earthquakes and tsunamis. The government, however is committed to continue with nuclear power programme, and is trying its best to achieve the target of 20 GWe by the year 2020. However, there has been opposition from the environmentalists and anti-nuclear and other civil society groups who have been accused by the prime minister of being manipulated by foreign NGOs opposed to nuclear energy development in India. These NGOs are being charged for illegally diverting foreign funds for promoting the anti-nuclear protests in the Kudankulam and Jaitapur. The anti-nuclear groups have been holding up the commissioning of the Kudankulam reactors. Similarly they are protesting against the Jaitapur power plants on the Arabian Sea coast in Maharashtra, where six power plants with a total capacity of 9900 MW (e) are being planned. Kudankulam 1 reactor has been commissioned in September 2013.

Germany: The Europeans as usual have reacted strongly and Germany has virtually stopped any further expansion of its civilian nuclear power programme. It has temporarily shut down seven of its 17 nuclear plants and reversed the pro-nuclear policy adopted last year. It imposed a three month moratorium on the life time extension. It also ordered general safety reviews of all operating plants. Chancellor Angela Merkel ordered the seven nuclear reactors, which began operation in 1980 or earlier, to be immediately shut down. The addition of another unit, already in long term shutdown, resulted in making a total of 8336 MWe go offline, which was about 6.4 per cent of the country's generating capacity. With the shutting down of these 8 reactors the share of nuclear power in the country's energy mix came down 27.3 per cent in March 2012 from 28.4 per cent in 2010.[33] In view of the mounting anti-nuclear pressure the government on May 30, 2011, decided to revive previous government's phase-out plan and close all reactors by 2022.The decision was confirmed by the German parliament Bundestag on July 8, 2011. The Parliament also approved the construction of new coal and gas fired plants despite retaining its CO2 emission reduction targets. This meant that the eight oldest reactors will remain closed and the remaining nine will close by the end of 2022. In order to meet the energy deficit, Germany will increase its electricity imports from France, Poland and Russia, generated mainly from nuclear sources. It would also mean a significant increase in its import of gas from Russia.[34] Since Germany would be importing nuclear power from the neighbouring countries its opposition to nuclear energy does not sound very convincing and smacks of hypocrisy.

France: France has a network of 58 nuclear reactors with a net 63,163 MWe (63.163 GWe) generating capacity. It gets about 74 per cent of its electricity from nuclear power. Immediately after the Fukushima nuclear accident in March 2011, France announced that it had no intention to turn its back on nuclear power and reiterated the significance of the nuclear power. However, it decided to undertake a general safety review and a check up of all operating plants with a focus on flooding risks, earthquakes, loss of cooling system and operational management crisis. The French Nuclear Safety Authority (ASN) found no reason to close down any nuclear facility.[35]

United Kingdom: Of the 17 operating reactors all but one reactor would retire by 2023. Nuclear power would continue to be a part of energy mix of the country. UK also decided to go for a general safety review and check up of all the operating plants immediately after the Fukushima accident. It called for a report on the implications of the situation in Japan for the UK nuclear power industry and the lessons to be learned. As of March 2012 UK has 17 units in operation contributing about 18 per cent to the electricity generation in the country. By 2018 some of the first of the 19 GWe new generation nuclear plants are expected to be online. UK has put in place a very thorough assessment process for new reactor design and sitting.[36] Mike Wightman, HM Chief Inspector of Nuclear Installations, in an interim report submitted on May 18, 2011 on the 'Japanese Earthquake and Tsunami: Implications for the Nuclear Industry', clearly stated that there was no reason to curtail the operation of nuclear power plants or other nuclear facilities in the UK. This conclusion was reached after studying the direct causes of the Fukushima accident. Not only this, but at the same time, in response to the Fukushima nuclear accident, the UK nuclear power industry reacted responsibly and appropriately and took the lead to promote a strong safety culture.

The report also referred to the government's proposal to create an office for nuclear regulation with the objective of enhancing the public confidence in the UK's nuclear regulatory regime and to effectively face any challenges in the future. No weaknesses were detected in the UK nuclear licensing regime. The fears of flooding were also brushed aside. In case, there were any possibilities of flooding, necessary changes would be carried out in the plant layout. Consequently there was no need to change the present sitting strategies for new power stations; and there would also be no departure from multi-plant sites.[37]

South Korea: Like many other prominent countries that are dependent on nuclear power, South Korea too reacted cautiously to the Fukushima nuclear accident. It emphasised the need for the safety and safety oriented reactor designs while reiterating its commitment to nuclear power. According to the IAEA as of March 2012, South Korea had 23 operating reactors with a total nuclear capacity of 20,671 MWe or 20.671GWe. It gets about 32.2

per cent of its electricity from nuclear power[38]. South Korea has 9 reactors under construction, on order or planned. The first two became operational in 2008 and one in 2009. The construction on the remaining four is scheduled to start between 2012-15.[39] By 2015 it would have seven reactors under construction. South Korea reaffirmed its aim to generate 59 per cent of its electricity from nuclear power by 2030. Declaring nuclear energy to be a strategic priority, it has planned to increase its capacity by 56 per cent to 27.3 GWe by 2020 and 43 GWe by 2030. It has also emerged as an exporter of nuclear reactors and has signed a deal worth $200 billion with UAE to supply four nuclear reactors.[40]

The South Korean president Lee Myung-bak has emphasised Korea's nuclear safety and emergency planning in response to the Fukushima nuclear accident. The need to improve safety of nuclear power plants in earthquake and tsunami conditions was emphasised by the members of the South Korean Congress in discussions with the South Korean deputy minister of education, science and technology.[41] According to Lee Jin-ho, Director International Cooperation at the Korean Institute of Nuclear Safety, Korea was engaged in doing "stress tests and checking safety. We are doing all kinds of evaluation and experiments". The government is "committed to civilian nuclear power and there is no change in Korea's nuclear energy policy" according to nuclear industry department of the government. However, for the time being it has put its sales pitches to countries such as Egypt and Malaysia on hold.

UAE: The UAE which has contracted to import four nuclear reactors from South Korea has decided to continue with its nuclear energy programme and wants it to contribute about 25 percent to its energy mix and expects to start its first reactor by 2017. However, UAE's Federal Authority for Nuclear Regulation (FANR) has asked the Emirates Nuclear Energy Corporation (ENEC) to explain how it proposes to apply the lessons of Japan's nuclear crisis in the context of UAE's nuclear programme.[42]

Taiwan: The Taiwanese government has been unnerved by the nuclear accident in Fukushima. Its President Ma Ying-jeou, after review of the safety of nuclear reactors and level of emergency response preparedness for nuclear accidents, announced that his government would not suspend the ongoing construction of its fourth nuclear power plant but would continue to upgrade nuclear plant safety and its capacity to face and resist disasters. He also declared that the operations of Plant 2 and 3 too would not be suspended. It was pointed out by the President that Taiwanese reactors incorporated advanced and safer fourth –generation technology and were less likely to be hit by a tsunami. Moreover, they were built on hill sides about 12-15 meters above ground level. He, however, said that Taiwan would spend more money to install a self-developed on-site earthquake warning system which would provide timely warning of 10 seconds before a quake struck the island.[43]

Indonesia: On November 30, 2011, Indonesia's state enterprise minister, Dahlan Iskan, announced the approval of plans for construction of its first nuclear power plant in the Tan Jung Ultar Muntok Cape region, West Bangka. Earlier in October Hudi Hastowo, head of Indonesia's National Nuclear Energy Agency had announced that the nuclear plant had been delayed because of the Fukushima nuclear accident. Despite the controversy around the power plant, the minister said that Indonesia would proceed with the project since the radiation leak at the Fukushima did not claim any lives. Indonesia has plans to build two nuclear power plants in Bangka-Belitung province, off the southern coast of Sumatra with start-up dates in 2021 and 2022.[44]

Indonesia is situated within the "Ring of Fire" which makes it vulnerable to volcanic eruptions, earthquakes and tsunamis. Despite the risks and concerns raised by the nuclear crisis in Japan it has declared that it still plans to build nuclear reactors. According to a spokesperson for the Indonesian National Nuclear Energy Agency," we have to be more concerned about the needs of this country for energy in the near future" and for that "we will build nuclear reactors outside the earthquake zone and in accordance with the IAEA guidelines". Indonesia is planning to build a nuclear reactor in Muria in Java by 2016, which is considered the safest place in Java, as it is quite far from the fault line, and volcanoes. Indonesia is also studying the feasibility of building a nuclear plant on Bangka Island off the east coast of Sumatra. It would have underwater cables to carry electricity to Java. Indonesia's policy to pursue nuclear power programme is based on its energy requirements as it is one of the most populous country in the world. It requires all types of power to meet its growing energy needs, according to Ferhart Aziz, the spokesperson for National Nuclear Energy Agency. Indonesia already safely operates three nuclear research facilities and has assured that all safety measures will be taken for building nuclear power plants.[45] However, some critics of Indonesian nuclear plans have argued that Indonesia has no "business locating them atop the Ring of Fire, that arc of fault lines and volcanoes encircling the Pacific Basin."[46] Indonesia signed the CTBT and deposited the instrument of ratification to the UN on February 7, 2012. Many ASEAN counties have not signalled their complete commitment to international safety norms. Indonesia is the only ASEAN country which has signed all the three major United Nations conventions on nuclear safety.[47]

Indonesia's commitment to the international safety and security agreements can be linked with its civilian nuclear energy development plans. Indonesia's hesitation regarding the construction of nuclear power plants has been more political rather than technical in nature as Indonesia has been extremely cautious that its energy development programme should be in compliance with international norms and it should have a well developed system to protect and control nuclear and radiological sources. It has seen to it that its civil nuclear industry was completely in accordance with the well established international norms for the physical protection of nuclear materials which has been vouched for by the IAEA's International Physical Protection Advisory Services (IPPAS) in 2001. It has also been conscious of the fact that it should "launch its nuclear programme only when its nuclear infrastructure was able to sustain the demands of the nuclear industry."[48]

Malaysia: Malaysia, Philippines and Thailand have delayed their nuclear programmes though they have not cancelled their plans entirely. Malaysia has a well established nuclear research and education infrastructure. The Malaysian Institute for Nuclear Technology Research (MINT) has been operating the MW Triga research reactor since 1982. MINT was renamed as the Malaysian Nuclear Agency in 2007 to promote peaceful uses of nuclear energy with the objective of reducing reliance on natural gas which has an almost two third share in electricity generation in the country. It was in 2009 that Malaysia undertook a comprehensive energy policy study and accordingly the government decided to include nuclear power in the overall energy mix of the country. A Nuclear Power Development Steering Committee was established to plan and coordinate the nuclear power development programme through three working groups.[49]

In a seminar on nuclear energy and the lessons of Fukushima organised by the Malaysian Bar, Dr Aishah Bidin, Dean Faculty of Law, University of Kebangasaan, Malaysia touched upon lack of a legal and regulatory framework for nuclear power in Malaysia and argued that it has yet to update its laws to comply with the IAEA's regulatory framework. The Malaysian

Act 304 does not provide for emergency response preparedness to ensure that Malaysia can respond to something like the Fukushima nuclear accident. There is thus a strong need to strengthen the legal infrastructure for nuclear activities in Malaysia. However, the revised Act 304 has additional provisions and the seven new heads now cover safety, nuclear security, safeguards, decommissioning, emergency preparedness and response, enforcement, and offences. It also has a provision for establishing a radioactive waste fund. At the same time Malaysia has yet to ratify the 1994 Convention on Nuclear Safety. Malaysia has signed but not yet ratified the 1963 Vienna Convention on Civil Liability for Nuclear Damages.[50]

Malaysia will have to address the legal issue of indemnifying the victims of a nuclear accident—whether the government would be liable if the nuclear power plant operator could not pay. There are a host of questions regarding liability that need to be addressed. The lessons from Fukushima could be of help to the government. The existing provisions relating to liability are given in Section 59 of the Act 304 which fixes the maximum compensation at RM 50 million.[51]

There was overwhelming public acceptance of nuclear energy before the Fukushima nuclear accident. Almost 79 per cent favoured the use of nuclear energy, out which almost 42.9 per cent were strongly in favour, while 37.7 per cent were somewhat in favour. About 52 per cent attributed a significant role to the nuclear energy in the country's energy mix.[52] However, after the Fukushima nuclear crisis there is greater public concern regarding the safety of nuclear power and many state governments have said 'no' to having power plants in their backyards. At the same time there is a growing awareness about the radiation issues among the general public and most of the discussions, dialogues, debates revolve around nuclear technology and safety related issues. Despite these shifts there is no change in the government policy and nuclear energy remains an important option for the country with the first nuclear power plants targeted to be operationalised by 2021.[53]

The Malaysia Nuclear Power Corporation (MNPC), incorporated in January 2011, has been entrusted with the task of making Malaysia nuclear ready by 2013 According to the Economic Transformation Programme, the timeline for generating nuclear power was 2021 and the target after the Fukushima nuclear crisis remains the same. Malaysia needs it for electricity generation, research and development, submarines and its space programme.

However this will require: public acceptance, ratification of all the international treaties, establishment of a transparent regulatory framework, and intensified public awareness campaigns.

Vietnam: Unlike Thailand, Philippines and Indonesia, Vietnam was absolutely unequivocal about its commitment to the nuclear power programme. The Vietnamese Agency for Radiation and Nuclear Safety and Control (VARNSAC) asserted that it would not review its nuclear power development programme despite the nuclear crisis in Japan. Instead of affecting its nuclear programme Vietnam would learn serious lessons from the Fukushima crisis. The head of the VARANSAC, Ngo Dang Nhan, assured that his country would choose the most advanced technology to ensure that its nuclear power is safe. Vietnam would guarantee maximum safety for its future nuclear power plants though the chances of Vietnam being affected by powerful quakes was quite low. Nhan also said that Vietnam would also develop human resources to carry out the nuclear energy development programme and also adopt a "sound legal framework for developing, operating and maintaining nuclear power plants."[54] In 2010 Vietnam announced its plans to build as many as 13 nuclear power stations with a capacity totalling 16,000 megawatts over the next two decades.[55]

It may be mentioned here that in a meeting with his Vietnamese counterpart on October 31, 2010, the Japanese Prime Minister Naoto Kan agreed that Japan would construct two nuclear power plants in Vietnam and they would sign bilateral nuclear cooperation agreement, thus "paving the way for export of power generation technologies to the fast growing economy". The agreement was the result of a series of summit meetings involving the ASEAN, of which Vietnam is a member and Japan a dialogue partner.[56] The Vietnamese Prime Minister Nguyen Tan Dung on his part assured the Japanese government that his country would stand by its decision to buy two reactors despite the Fukushima nuclear disaster. It was clear that Vietnam still had faith in the Japanese nuclear technology. He also called for continued cooperation with Tokyo for Vietnamese nuclear power programme.[57]

The decision of the Japanese government to export nuclear reactors at a time when there was no clarity about the future of nuclear energy within the country, and the prime minister was talking in terms of reducing the country's reliance on nuclear power, in response to people's growing anti-nuclear feelings, has been criticised as a policy "full of contradictions", "unprincipled" and "hypocritical" as far as international community was concerned.[58] It was said to be misguided and the people were likely to "look at its energy policy with suspicion"[59]

Vietnam-Russia Nuclear Cooperation: In October 2010 Vietnam signed an agreement with Russia as per which Atomstroyexport would build the Ninh Thuan 1 nuclear power plant and would construct two light water reactors VVER-1000 or VVER-1200 . It would be a turnkey project and would become operational from 2020 and would be financed by the Russia's ministry of finance to the tune of 85 per cent for the first reactor, provide the fuel and take back the spent fuel, for the life of the plant. Consequently a $9 billion agreement was signed in November 2011, for the construction of its first reactor and a second agreement covered the establishment of a nuclear science and technology centre.[60] The lending period was expected to be 28 years. Russia also agreed to fund the plant's feasibility study to be conducted by the E4 Group consisting of OJSC, Energo Project Kiev Co, and Energo Project Technology Co., and the state utility EVN. The Japan Atomic Power Co also signed a contract to carry out an 18-month feasibility study for the NinhThuan 2 power plant. Vietnam was also negotiating an agreement with US on civilian nuclear technology. Vietnam welcomed overseas assistance. This was a sort of signal to American companies such as General Electric Co to compete for contracts with Russian, Chinese, Japanese and French companies.

The Vietnamese President Truong Tan Sang and the South Korean President Lee Myung-Bak in a joint statement issued after talks in Seoul agreed that Vietnam would seek greater cooperation with South Korea for a nuclear power plant using South Korean Technology. The French development agency too has pledged its support to Vietnam's nuclear endeavours. Vietnamese officials met with the representatives of AREVA to benefit from their "experience in building and operating nuclear power plants: and also for ascertaining AREVA's capabilities for 'providing to fuel rods, converting and enriching uranium and reprocessing and handling nuclear waste".[61] Vietnam has also reportedly entered into a nuclear cooperation agreement with India, South Korea, Canada and Argentina.

Conclusion

Post-Fukushima trends with regard to nuclear energy are mixed. The nuclear accident to a certain extent has adversely affected the projected growth of nuclear energy. It has boosted the anti-nuclear movement raised concerns among the public about the dangers of the nuclear energy in terms of radiation, environmental damage and public health. However, except for a few European countries, like Germany, which has decided to dismantle its eight reactors and phase out nuclear energy completely from its energy mix, the commitment of most governments particularly in the developing countries to nuclear energy as a clean energy source and a viable option for energy security remains intact. There is a consensus amongst all the countries that better safety measures should be adopted for operating plants and new safety designs should be developed for future nuclear reactors. The governments will also have to work to ensure public acceptance and enhance awareness about the positive aspects of nuclear energy and nuclear power as a safe energy source.

India will have to ensure nuclear reactor fuel security particularly uranium mining and import of nuclear fuel. India has made significant progress in indigenous nuclear reactor technology development. The DAE is engaged in developing new bigger reactors and is also involved in research for developing the fusion reactor and fusion energy. However, to achieve all this India needs to allocate more money for R&D. India has rightly decided to import European Pressurised Reactors (EPR) from France. The EPR technology, though considered an untested technology, is a core edged technology and already four reactors are under construction in Finland, China and France. There are no safety problems as the best safety measures have been incorporated in the reactor design. India has also signed nuclear cooperation agreements with Russia, France, South Korea, US, and Canada. Negotiations for cooperation with Japan are at an advance stage. Talks are also in progress with Argentina, Brazil, and Australia.

The experience of China, South Korea and French, can be instructive to India. The way South Korean nuclear industry has expanded in a short span leading to the emergence of Seoul as a major exporter of nuclear power reactors can be a sort of lesson for India as it too aspires to be an exporter of nuclear reactors in the near future. India can also learn from the major civilian nuclear power countries such as Japan, South Korea, France, Russia and China. India has the potential to emerge as an exporter of nuclear reactors to the emerging nuclear markets particularly the small and medium type reactors. For India to venture in to the international nuclear export markets requires it to study how the major nuclear power countries have managed finances for this sector. It could raise finances by issuing of nuclear power bonds, and through public private Partnership.

Safety standards at India's nuclear reactors are of international level. These can become a major incentive for countries to import nuclear reactors from India. In addition to the prevailing nuclear safety standards the Indian government in the light of the Fukushima nuclear accident has ordered a further review of safety standards. The Atomic Energy Regulatory Agency will play a crucial role in implementing and enforcing of the safety standards. The Indian government has taken steps to meet the domestic challenges constraining India's nuclear power development plans such as: (i) opposition and resistance from the civil society, (ii) dispelling misperceptions about nuclear power and projection of the usefulness of the nuclear energy, (iii) carrying out institutional changes i.e. decentralisation of the nuclear energy management structures, transparency and accountability, (iv) managing protests against the uranium mining and the opposition to the nuclear power projects at Jaitapur and Haripur, etc. by environmental groups and anti-nuclear NGOs. India's nuclear cooperation with a number of countries and its own indigenous nuclear power plant development would invariably raise the issue of nuclear liability in case of nuclear accidents. It would entail a strong new institutional legal framework to deal with such eventualities. This now has been addressed by the nuclear liability bill passed by the Indian Parliament in 2011. It has been discussed in a separate section of the book.

In view of the challenges and constraints, the most troublesome question which needs to be addressed is whether India would be able to achieve the projected target of 20 GWe and 63GWe nuclear energy by the end of 2020 and 2032 respectively. It is hard not to be pessimistic considering the slow pace at which the power plants are being built. These targets now have been revised and it is planned to reach 10,080 MW by 2017 and 27,480 MW by 2023-24.[62] The Japanese government allowed the restarting of the two reactors, followed by restarting of more reactors. Prime Minister Shinjo Abe and his predecessor Prime Minister Noda are convinced that there is no alternative to nuclear energy if Japan has to retain its present standard of living and sustain a reasonable economic growth. However, the majority of people are still opposed to the nuclear energy. Japan is still debating the future of the nuclear energy so far there is no consensus about its future energy plans.

NOTES & REFERENCES

1. Veronika Oleksyn (2011), "Crucial lessons in reactor crisis cited," *The Boston Globe,* April 5, 2011, URL: http://www.boston.com/news/world/asia/articles/2011/04/05/crucial_lessons_in_reactor_crisis_ cited/, accessed on March 27, 2013.
2. Martin Fackler and Andrew Pollack (2011), "Japan scraps plan to build new reactors; prime minister shelves reactor plans and vows new focus on renewables", *International Herald Tribune,* May 11, 2011, URL: www.nytimes.com/2011/05/11/asia/japan.html, accessed on March 9, 2013.
3. "Nuclear Energy', *International Herald Tribune - New York Times,* March 9, 2012.
4. "Japan-Earthquake, Tsunami and Nuclear Crisis 2011", *International Herald Tribune - New York Times,* February 27, 2012.
5. "Nuclear Energy - 2011, Japan Nuclear Crisis; Overview", *International Herald Tribune,* March 8, 2012.
6. Chico Harlan (2011), "Noda weighs public's nuclear firms' export ambitions", *The Japan Times,* November 23, 2011, URL: http://www.japantimes.co.jp/news/2011/11/23/national/noda-weighs-publics-nuclear-fears-firms-export-ambitions/#.UzKM2aiSxIw, accessed on March 10, 2012.
7. Ibid. Chico Harlan.
8. Ibid Chico Harlan.
9. "Diet Ok's new agency to run Tohoku post disaster reconstruction efforts", *The Japan Times,* December 10, 2011.
10. "Nuclear accords with four nations near Diet passage", *The Japan Times,* December 1, 2011.
11. "Japan to start two reactors to meet electricity needs", *The Extinction Protocol,* June 17, 2012, URL: http://theextinctionprotocol.wordpress.com/2012/06/17/japan-to-start-two-reactors-to-meet-electricity-needs/, accessed on July 7, 2013.
12. "Japan regains nuclear power as first reactor resumes operations", *Reuters,* July 05, 2012, URL: http://in.reuters.com/article/2012/07/05/us-japan-nuclear-reactor-idUSBRE86400T20120705, accessed on July 7, 2013.
13. In a May 2012 poll conducted by Japanese daily *Asahi Shimbun,* 54% respondents opposed the restart of the reactors. Other polls showed even higher opposition - almost 71%. See Krista Mahr (2012), "Start Your Engines: Japan Gets Ready to Return to Nuclear Power", *Time World,* June 15, 2012, URL: http://world.time.com/2012/06/15/start-your-engines-japan-gets-ready-to-return-to-nuclear-power/, accessed on July 7, 2012.
14. Ibid, Krista Mahr.
15. International Atomic Energy Agency, *Nuclear Power Plants Information - Power Reactor Information System (PRIS),* URL: http://www.iaea.org/pris/Home.aspx., accessed on July 7, 2013.
16. European Nuclear Society, *Nuclear power plants worldwide,* URL: www.europeannuclear.org/ifo/encyclopedia/n/nuclear-power-plants-world-wide.htm, accessed on March 12, 2013.
17. World Nuclear Association, *World Energy Needs and Nuclear Power Updated on December 2011..*
18. Ibid. *World Energy Needs and Nuclear Power.*
19. World Nuclear Association, *Nuclear Share figures 2000-2010 updated as on June 2011.*
20. Benjamin Haas (2011), "Japan Crisis probably won't end China's commitment to nuclear power", *Los Angeles Times,* March 19, 2011, URL: http://articles.latimes.com/2011/mar/

19/business/la-fi-china-nuclear-20110319, accessed on July 28, 2013.

21. Ibid, *World Energy Needs and Nuclear Power.*
22. European Commission (EC), *Energy Roadmap 2050*, Brussels (COM (2011) 885/2, URL: http://ec.europa.eu/energy/publications/doc/2012_energy_roadmap_2050_en.pdf, accessed on March 13, 2013. Also see *World Nuclear Association*, "World Energy Needs and Nuclear Power" December 2011.
23. Mathew Bunn and Martin B. Malin (2009), "Enabling a Nuclear Revival-And Managing its Risks, *Innovations (MIT)*, 4 (4): 173-191.
24. Ibid, Mathew Bunn and Martin B. Malin (2009), pp. 173-191.
25. Ibid, Mathew Bunn and Martin B. Malin (2009), pp. 173-191.
26. Report prepared by an independent Commission at the request of the Director General of the International Atomic Energy Agency, "*Reinforcing the Global Nuclear Order for Peace and prosperity: The Role of IAEA to 2020 and Beyond*", IAEA, 2008. Gov/2008/22-GC(52)INF/4, May 23, 2008.
27. World Nuclear Association (2012), *Plans for New Reactors World Wide updated on February 2012*, URL: www.wold-nuclear.org/info/inf17html, accessed on March 9, 2013.
28. Kevin Tu (2011), "Nuclear Crisis in Japan: Preliminary Policy Implications for China", *Carnegie Endowment for International Peace*, April 1, 2011.
29. "Japan Disaster - Fukushima crisis spotlights Chinese nuclear concerns", *Taipei Times*, March 18, 2011, URL: http://www.taipeitimes.com/News/world/archives/2011/03/18/2003498505, accessed on March 10, 2013.
30. Staff Reporter, "China to focus on solar power after Japan's nuclear crisis", *Want China Times*, April 1, 2011, URL: http://www.wantchinatimes.com/news-print-cnt.aspx?cid=1105&M, accessed on March 14, 2013.
31. Zhao Qizheng, a spokesperson for China's political advisory body said, "We will develop nuclear power under "extreme safe" preconditions after drawing lessons from the Fukushima nuclear accident". The honorary President of State Grid Electric Power Research Institute and a deputy in China's top legislature, Xue Yusheng, said, "The risk factor must be taken into account at the very beginning of a plan to build a nuclear power station, because potential risks at nuclear power stations are inevitable". He further said, "The development of nuclear power matters to national welfare and the people's livelihood. Thus, we cannot cope with plans that view nuclear disaster only as small probability and without preparations measures". According to him any kind of power generation has its own disadvantages, and, for the sake of energy security, China, a big energy consumer should not give up nuclear power. To ensure energy security, we have to take some potential risks and should not avoid use of nuclear power. See "China rethinks nuclear power one year after crisis in Japan", *Xinhua - China Daily*, March 13, 2012, URL: www.chinadaily.com.cn/bizchina/2012-03/content-14822402html, accessed on March 14, 2013.
32. Ibid
33. Nuclear Energy Institute, *Nuclear Energy around the World*, URL: http://www.nei.org/resourcesand statistics/nuclear_stats/world statistics1, accessed on March 14, 2012.
34. World Nuclear Association, *Nuclear power in Germany – Updated Till February 2012*, URL: http://www.world-nuclear.org/info/Country-Profiles/Countries-G-N/Germany/, accessed on March 17, 2013
35. Areva (2011), *Impact of Fukushima event on nuclear power sector: preliminary Assessment*, March 25, 2011, URL: http://groupes.sortirdunucleaire.org/IMG/pdf/DocinterneAREVA-25Mars2011.pdf, accessed on March 14, 2013
36. International Atomic Energy Agency, *World nuclear Power Generation and Capacity*, URL: http://www.iaea.org/programmes/a2/index.html, accessed on March 18, 2013
37. Mike Weightman (2011), *Japanese Earthquake and Tsunami: Implications for UK nuclear*

Industry, Interim Report, May 18, 2011, URL: www.hse.gov.uk/nuclear/fukushima/interim-report.pdf, accessed on March 15, 2013.

38. International Atomic Energy Agency, *World Nuclear Power Generation and Capacity*, March 2012, URL: http://www.iaea.org./programmes/a2/index.html, accessed on March 15, 2012. According to World Nuclear Association nuclear electricity share is 31% as of February 22, 2012.
39. World Nuclear Association, *Nuclear power in South Korea*, February 22, 2012, URL: www.world.nuclear.org.info/inf81.html, accessed on March 16, 2012.
40. Ibid, "Nuclear power in South Korea".
41. Sungyed Choi (2011), "The Global Future of Nuclear Power after Fukushima", *Belfer Centre for Science and International Affairs*, March 16, 2011, URL: http://belfercenter.ksg.harvard.edu/power/2011/03/16/the-global-future-of-nuclear-po, accessed on, April 9, 2011.
42. "UAE planning first reactors, wants to learn from Japan crisis", *The Jordan Times*, April 8, 2011.
43. "Taiwan goes ahead with building 4th nuclear plant," *Asiaone*, URL: http://news.asiaone.com/News/Latest+News/Asia/Story/A1Story20110316-268373.html, accessed on April 10, 2013.
44. "Indonesian government approves nuclear construction plan", *Nuclear Engineering International,* December 5, 2011, URL: *www.neimagzine.com/*story.asp?StoryCode=2061327, accessed on April 10, 2012.
45. Brian Padden, "Despite Japan Crisis, Indonesia Pursues Nuclear Power", *Voice of America*, URL: http://www.voanews.com/content/despite-japan-crisis-indonesia-pursues—nuclear-power-117924389/136456.html, accessed on April 10, 2012.
46. William Pesek (2011), "Indonesia's nuclear ambitions, Post-Fukushima", *TODAY online*, March 16, 2011, URL: http://www.todayonline.com/Print/ommentary/EDCI110321-0000201/Indonesia-nu, accessed on April 10, 2012.
47. Adiwardojo (2011), *"Nuclear Infrastructure Development in Indonesia"*, Presented at IAEA Technical Meeting on the Introduction of Nuclear Power Programme: Management and Evaluation of a National Nuclear Infrastructure, 8-11 February 2011, IAEA, Vienna, URL: http://www.iaea.org/NuclearPower/Downloadable/Meetings/2011/2011-02-TM-WS-Vienna/Day- 1/Adiwardojo_Indonesia.pdf, accessed on April 10, 2013. It is a party to all the international agreements relating to nuclear safety and security including Convention on the Physical Protection of Nuclear Materials; the Joint Convention on the Safety of Spent Fuel Management and on the Safety of Radioactive Waste management; and the Convention on Nuclear Safety; Convention Assistance in the Case of a Nuclear Accident or Radiological Emergency; and Supplementary Compensation for Nuclear Damages.
48. Ibid, Adiwardojo, "*Nuclear Infrastructure Development in Indonesia*".
49. World Nuclear Association (2012), *Emerging Nuclear Energy Countries updated on March 2012.* These three are, Nuclear Power Program Working Group under the Malaysian Nuclear Energy Agency (MNA); the Nuclear power Project Working Group under the utility Tenaga Nasional Berhad (TNB); and a Legal and Regulatory Coordination Working committee involving the Atomic Energy Licencing Board (AELB) and the Energy Commission. The steering Committee's Nuclear Power Infrastructure Development Plan (NPIDP) was given the target date of 2013 at which time the government would decide whether to proceed with it or not.
50. Kiu Jia Yaw (2011), "Seminar on Nuclear Energy: Lessons from Fukushima", *The Malaysian Bar*, November 14, 2011, URL: http://www.malaysianbar.org.my/environment_and_climate_change/seminar_on_nuclear_energy_lessons_from_fukushima_14_nov_2011.html, accessed April 12, 2013.
51. Ibid, Kiu Jia Yaw.
52. Mohamad Puad Haji Abu (2011), "*The Malaysia nuclear Energy Option: A Post Fukushima*

Perspective", 2nd Technical Working Group Meeting on Nuclear power Infrastructure", Vienna, Austria, 23-26 May 2011 cited in Mazleha Maskin, "Update on Nuclear Energy Programme in Malaysia", presentation made at the Interregional Workshop on Advanced Nuclear Reactor Technology for Near – Term Development, Vienna Austria, 4-8 July 2011.

53. Ibid, Mohamad Puad Haji Abu.
54. Van Oanh (2011), "Vietnam nuclear power project to move on despite Japan Crisis", *Saigon Times*, April 9, 2011.
55. "Russia to lend Vietnam $9 billion for first nuclear plant", *Bloomberg*, November 25, 2011, URL: http://www.bloomberg.com/news/2011-11-22/russia-to-lend-vietnam-9-billion-for-first-nuclear-plant.html, accessed on March 22, 2012.
56. "Japan, Vietnam tie up on reactors", *The Japan Times*, November 1, 2010.
57. "Vietnam to stick buying two reactors from Japan: Hatoyama", *The Japan Times*, March 21, 2012.
58. Editorial, "Unprincipled nuclear power policy", *The Japan Times*, December 16, 2011.
59. Editorial, "Nuclear Export Policy Misguided", *The Japan Times*, November 11, 2011.
60. World nuclear Association, "Nuclear Power in Vietnam - January 2012".
61. Pham Nghiem Xuan Bac and Luu Tien Ngoc, "Leading the Nuclear Charge-Vietnam's Quest to Develop Nuclear Power", *Vision and Associates*, URL: http://www.vision-associates.com/legal_news_detail/2/350/Leading-the-Nuclear-Charge-%E2%80%93-Vietnam%E2%80%99s-quest-to-develop-nuclear-power, accessed on April 2, 2012.
62. Press Information Bureau (2012), "Nuclear Power Target", *Government of India*, August 9, 2012, URL: http://pib.nic.in/newsite/erelease.aspx, accessed on October 12, 2013.

CHAPTER 2

Relevance of Nuclear Energy in India's Energy Mix: Energy Security and Economic Development

K.D. Kapur

Introduction

With an average growth rate of 8 per cent Indian economy has been on a high growth rate curve for more than a decade. The rapid increase in GDP translates into higher energy consumption and the overall development of infrastructure. It is a stark reality that there are about 600 million Indians who still do not have access to electricity. At the same time for about 700 million people biomass is their primary energy resource for cooking.[1]

India's energy basket is primarily fossil fuel centric (coal, oil and gas). India is making efforts to increase the share of non-polluting energy sources such as renewable energy and nuclear energy. Though the share of nuclear power is currently very small it has the potential to increase its share in the country's energy basket. In India the demand for energy has outpaced supply leading to increased energy import dependence. This chapter discusses the various energy sources of India and the relevance of nuclear and renewable sources of energy for the country's economic development and its energy security.

Planning Commission of India's Twelfth Five Year plan (2012-17), has made projections for the energy requirements for this period on the assumption that the Indian economy will grow at the rate of 9 per cent per year. To fuel this rate of growth the energy supply is projected to grow at around 6.5 per cent per year for the next five years. To meet this demand India would have to expand its domestic production in critical energy sub sectors, particularly petroleum, gas and coal, and the balance requirement through imports. Because of the dominance of coal in India's energy sources, green house gas (GHGs) emissions have been rising and today India is one of the five largest GHGs emitters although its per capita emission remains far below the global average. This has also been responsible for increased carbon dioxide emissions, from 774 million tons in 1994, to 1050 million tons in 2003 marking an increase of 355 million tons from the base year emissions over a period of ten years. The sustained growth and development of the economy requires an assured and sustained supply of energy.

It is in this context that renewable sources of energy—including nuclear—are seen as one of the options for addressing the challenges of energy security along with meeting the commitment for reducing GHGs thereby preventing climate change and global warming. India's National Action Plan for Climate Change released in 2008 envisages that 15 per cent of our energy consumption will come from renewable sources of energy by 2020, a target which is quite ambitious but at the same essential and needs to be pursued vigorously.

The Jawaharlal Nehru National Solar Mission (JNNSM) is one of the eight missions launched for meeting the challenge of climate change. It aims to achieve a target of grid connected solar electricity production of 20,000 MW and off-grid production of 2,000 MW by the year 2022. It may be mentioned that globally solar power industry has been growing at quite a pace.[2] India's solar energy mission is showing positive trends and can have a definite impact on India's energy security and future energy requirements.

Nuclear energy being a non-polluting source of power, with very little carbon emissions, is also the focus of Indian government's future energy plans. The government is committed to increase its share and has set a target 20 GW nuclear power generation by 2020. Wind power, with a total contribution of about 18,420 MW as on December 31, 2012, is the most important component of India's renewable energy mix. Biomass too is making a considerable contribution towards the county's energy requirement particularly in the off-grid market. This chapter has been divided into following sections: Introduction; Energy security; Economic development and Environmental Concerns; Energy Scenarios; Fossil fuels; Non-fossil fuels;

Nuclear Power; Comparisons between different sources of energy; and Conclusion.

Energy Security, Economic Development and Environmental Concerns

Energy security related threats can be categorised into two: those "affecting the economic growth" of the country and those that are "unique to conditions of poverty". Threats and conditions that directly affect the economic growth and normal functioning of the economy can be identified as: supply constraints; costs; import dependence and geopolitical tensions. Non-access to energy for the marginal and peripheral sections of the society is a major component of the threat to energy security. Other threats are: energy inequity; and use of dirty and traditional biomass fuels associated with the health risks.[3]

The Planning Commission of India in its 'Integrated Energy Policy' has outlined the broad components and parameters of energy security holding that India would be an energy secure nation when it:...can supply lifeline energy to all (its) citizens irrespective of their ability to pay for it as well as meet their effective demand for safe and convenient energy to satisfy their various needs at competitive prices at all times and with a prescribed confidence level considering shocks and disruptions that can be reasonably expected. This is an improved definition of the 1999 World Energy Assessment which considered the energy security as, "the continuous availability of energy in the varied forms in sufficient quantities at reasonable prices".[4] The Planning Commission's definition is comprehensive and broader in scope and squarely covers the issues relating to economic growth and the availability of energy to the poorer segments of the society. The incorporation 'cost' denotes externalities such as local and global environmental costs.

There is a symbiotic relationship between the energy supply in a country, its economic development and its rate of growth. Higher growth rates would demand an increase in energy supply; and a higher energy supply would ensure a higher growth rate. The supply of clean energy to its entire people is essential for:

> ...inclusive growth, meet the millennium development goals and raise India's human development index which at present compares poorly with several countries which currently are below India's level of development.[5]

The policy of the government of India is to ensure energy security for all. An equally important dimension of the energy strategy is to meet the energy needs of the vulnerable sections of the country with safe, clean, and convenient energy at the least cost. However, this must be achieved in a "technically efficient, economically viable and environmentally sustainable manner using different fuels and forms of energy, conventional and non-conventional, as well as new and emerging energy sources"[6]

Environmental Concerns

Industrialisation and development in the developing countries would involve more use of energy, increased resource utilisation, access to modern technology and enhanced production of consumer products. All this would have an impact on the environment requiring a balance between environmental protection and economic development. The policy makers have to strike a balance between the demands of the climate change and the developmental needs of the country.[7] Fossil fuels and other anthropogenic factors are held responsible for causing climate change which in turn can pose serious developmental challenges particularly to more vulnerable economies of the world. Though energy security and climate change are treated as distinct, but now it is accepted that such an approach would be suboptimal, so it would be appropriate to consider environmental threats as energy security threats.[8] It is argued that an over emphasis on the climate change, particularly in the case of developing countries, including India, should not underplay the more pressing developmental issues and other challenges such as low energy access, poverty and disease as their alleviation would definitely require more energy so as to achieve developmental goals.

Thus, though thermal power plants can help increase electricity generation; their environmental costs in terms of green house gases have proved them to be highly unsustainable. They have become the biggest health hazards particularly leading to increase in the respiratory diseases. The issue of the environmental cost of electricity generation would continue to be on the forefront of any debate regarding a country's energy mix. So any energy source which is less polluting is welcome. In this sense nuclear power, perhaps, is the best option, as the GHG emissions in its case are minimum as compared to all other sources of electricity generation.

Challenges to India's Energy Security

India faces a numbers of challenges in the area of energy security which should be addressed with all earnestness if the country is to continue its accelerated growth rate and meet the developmental needs of its population. Following are some of the important challenges.

Match Demand and Supply: Growing Dependence on Import of Energy

The accelerated economic growth of the economy involving higher energy consumption has created a mismatch between demand and supply resulting in dependence on energy import. The share of the imports of the Total Primary Commercial Energy Supply (TPCES) in India has gone up from 17.85 per cent in 1991 to 30 per cent in 2004-05. India's energy imports consisting of oil, gas and even coal are growing substantially and are likely to rise further in the coming years. India imports almost 76 per cent of its total oil consumption.[9] India's import of coal is also likely to increase considerably because of the problem in coal mining and the inferior quality of indigenous coal. India's import dependence between 2010-11 and 2016-17 on oil is expected to rise from 76 per cent to 80 per cent; natural gas from 19 to 28.4 per cent; coal from 19.8 per cent to about 22.1 per cent. In view of the globally rising energy prices imports would become more expensive[10]. India is therefore investing abroad in oil, natural gas and coal fields.[11]

The Planning Commission projections for the fossil fuel requirements for the year 2030-31 are: 337 to 462 Mt of oil; 99 to 184 Mtoe of gas; and 602 to 954 Mtoe of coal. The problem is to ensure the uninterrupted supply of energy. The supply risk scenario has been summed up by the Planning Commission as:

> The threats to energy security arise not just from the uncertainty of availability and price of imported energy, but also from the possible disruptions or shortfalls in domestic production. Supply risk from domestic sources, such as from a strike call in Coal India or Railways also needs to be addressed.[12]

An equally important concern relates to the constant rise in oil prices leading to inflation which may lead to an economic slowdown with all the consequences. Moreover, despite the existence of the energy resources, technical failures may disrupt the supply of energy to some people because of faulty generators, the tripping of transmission lines or oil pipelines springing leaks.[13] A study 'National Energy Map for India: Technology Vision 2030' undertaken by TERI on behalf of the government of India, in 2009, made suggestions regarding the action to be taken to ensure optimal energy supply for end use services in five energy consuming sectors: agriculture, commercial, residential, industrial and transport. It has also made projections regarding the commercial energy requirements in various scenarios.[14] Even if the contribution of energy from the hydro, nuclear, and renewable sectors increases about six times during 2001-31, they can at best contribute only 4.5 per cent of the total commercial energy requirements. Thus, it is evident that the pressure on the three conventional energy sources i.e. coal, oil, and gas will continue for at least next few decades.[15] Of these, gas is preferred option for power generation as well as fertiliser production. The study however, has argued that future energy requirements can be reduced by adopting various measures. [16]

Energy Efficiency: The integrated energy planning approach favours reduction of energy intensity.[17] It was observed that even in the Business As Usual (BAU) scenario there was a trend of declining energy intensity. The study had shown a decrease of 23 per cent from 0.022kgoe (kilogram of oil equivalent)/rupee of GDP in 2001 to 0.017kgoe/rupee of GDP in 2031. This shows that even with a GDP growth rate of 8 per cent and if the government plans and policies materialise as planned, the Indian economy was expected to be progressing along an energy-efficient path. The study also indicated that there was a scope for further reduction in energy intensity by adopting efficiency measures which could be decreased from 0.022kgoe/rupee of GDP in 2001 to 0.012kgoe/rupee of GDP in 2031.[18]

Protection of sea lanes: Since India is largely dependent on the import of energy, it would have to ensure that the sea lanes through which these supplies come to India are well protected. Presently India imports its oil, and gas from the Middle East, East Africa, Latin America, South East Asia and Oceania. Threats of piracy to the Indian energy trade along the Somalia coast have emerged as a major concern for energy supplies. The Indian navy now has been entrusted with the task of ensuring energy security. And to that extent it is being modernised to cope with its new role. Power projection by the Indian navy particularly in the Indian Ocean region is aimed at to protect assets in the southern Indian Ocean including the Diego Garcia Island. The Indian navy has provided security to Indian and foreign merchant ships and tankers carrying oil and gas by escorting them. Periodic naval exercises in the Indian Ocean region are conducted to ensure the security of sea lanes. India's Maritime Doctrine recognises that any disruption in the supply of oil from Persian Gulf could critically affect the country's interests and that "offshore oil and gas facilities" are "extremely vulnerable to disruptive attacks".[19]

Geopolitical risks to energy supplies: India's access to energy supplies could face geopolitical risks such as vulnerability of Indian energy assets in the turmoil ridden regions of Middle East (Iran, Egypt, Syria and Yemen). This could lead to a rise in oil prices, and disruptions in production and transportation; vulnerability of the proposed Trans-Afghanistan pipeline [Turkmenistan, Afghanistan, Pakistan and India (TAPI)] gas pipeline, to the perpetual Afghanistan—Pakistan unrest; the emerging competition between China and India for energy resources in different regions of the world, particularly the aggressive Chinese efforts to gain access to the oil and gas reserves in the Indian Ocean region by encircling India from East Africa to Burma and the Bay of Bengal by partnering the development of ports in some strategically important countries in the region which could be of concern to India. The recent objection of China to the Indian PSU entity Oil and Natural Gas Commission's (ONGC) oil exploration in some blocs in the Vietnamese waters in the South China Seas is a case in point. The global energy scenario in terms of demand and supply is also undergoing radical transformations. The US dependence on Middle Eastern oil is likely to be minimised in the coming years because of the massive production of shale gas in the country. The Arctic Sea in the coming years is expected to emerge as major source of hydrocarbons. The melting of the arctic ice and the opening of the northern sea route would shorten the distance between Europe and China. At the same time oil production from the OPEC is expected to go down considerably and the Latin America and Africa are likely to emerge as major producers of oil. Even the nature of demand is expected to change. The demand for oil in the OECD countries is going down. In the next couple of decades the developing countries, particularly China and India, would emerge as major importers of fossil fuel based energy.[20]

Domestic Challenges and Threats to Energy Supplies: A number of domestic factors have been identified as being the primary reasons for India's growing energy security concerns.[21] Energy supply concerns also emanate from the fact that India has only 0.8 per cent of known oil and gas reserves in the world and a huge population that is about 17 per cent of the world's population. At the same time, environmental concerns and the poor quality of coal and inaccuracies in reserve estimates has constrained it to avoid use of domestic coal and depend on imports of all categories of fossil fuels. There are also serious external security concerns arising from India's dependence on the import of crude oil which is expected to rise from 35 per cent in 2001 to 78 per cent by 2031. At the same time imports of natural gas and coal are also expected to grow. India's import dependence for oil, natural gas and coal would impel it to engage with energy rich counties more seriously.[22] Moreover, the demand for electricity and fuels such as LPG and CNG is also set to increase because of changed patterns of energy consumption. Important studies conducted by the TERI and the Planning Commission have come to the conclusion that in the next three to five decades India's economic development would be heavily dependent on fossil fuels. According to the Planning Commission's, Integrated Energy Policy report renewables may account for only 5 to 6 per cent of India's energy mix by 2031-32,[23] even if the country succeeds in developing 100,000 MW of renewable capacity over the next 25 years.[24]

Diplomatic Challenges to ensure energy security: India needs to activate its energy diplomacy. Various strategies have been suggested for securing uninterrupted supply of energy and cope with energy supply vulnerabilities. Some of them include the enhancing of maritime capacities and bolstering its naval power projection capabilities to its protect littoral area energy infrastructure; facilitate transfer of technology and investment in clean technologies; increase the share of renewable energy in the energy mix—wind, solar and hydroelectric; better protection of energy assets; and to obtain energy assets abroad. The ministry of external affairs (MEA) has established an energy security division with a view to collaborate in global energy diplomacy. India has signed a number of bilateral energy related Memorandums of Understanding (MOUs) and bilateral agreements with countries in Central Asia, Middle East, Africa and Latin America. India, however, still faces numerous challenges in its efforts to ensure its energy security. Indian energy security threats are primarily sea based except for the coal reserves which are located in the internal insurgency hit areas. The Indian navy needs more resources to ensure the security of the country's growing array of sea based energy assets.[25]

Pursuit of Energy Security: Reducing Dependence on Imports: In order to reduce dependence on the import of conventional energy fuels, it is recommended that the country should explore all possible options on the demand and supply side simultaneously, to reduce its total energy requirements as well as diversify the fuel resources mix. This could be achieved by taking an integrated approach to energy planning. TERI has recommended the following measures: (1) Reduce import dependence of coal by intensifying the exploration and production activity so as to increase the extractable coal reserves.[26] (2) The study recommends the adoption of the Benson boiler which operates at a higher steam properties pressure of 300 bars and at temperatures of 600 C. Moreover, they are highly cost effective if the power generation unit is of about 1000 MW capacity or more.[27] (3) The study also has recommended the adoption of efficient technologies such as: Integrated Gasification Combined Cycle (IGCC) technology; and refinery-residue based IGCC power generation plants.[28] (4) Reduction in the consumption of petroleum products. Enhance the share of public transport, Mass Rapid Transport System (MRTS), [29] and (5) Efforts should be made to make renewable energy resources competitive.

Energy Scenarios

As of 2010 coal based power generation constituted a share of 53 per cent in total energy production. Oil, gas, renewables and nuclear energy contributed 31, per cent 9 per cent, 6 per cent and 1 per cent respectively. By 2030 India is projected to become world's third largest energy consumer. According to Planning Commission to maintain an 8 per cent GDP growth through 2031-32, India's primary energy supply must increase three fold and electricity generation must go up as much as five times. India's indigenous sources of energy supply are not sufficient to meet the growing demand. Consequently, India is heavily dependent upon import of energy which at present amounts to almost two thirds of its consumption and is expected to go up to 90 per cent by 2030 as per the estimates of the International Energy Agency. By 2030 India's primary energy mix is likely to change slightly as coal, oil, gas, renewable and nuclear would contribute 46 per cent, 26 per cent, 9 per cent, 17 per cent and 2 per cent respectively. The oil component will drop by a few percentage points.

Figure 2.1: India's Primary Energy Mix, 2010

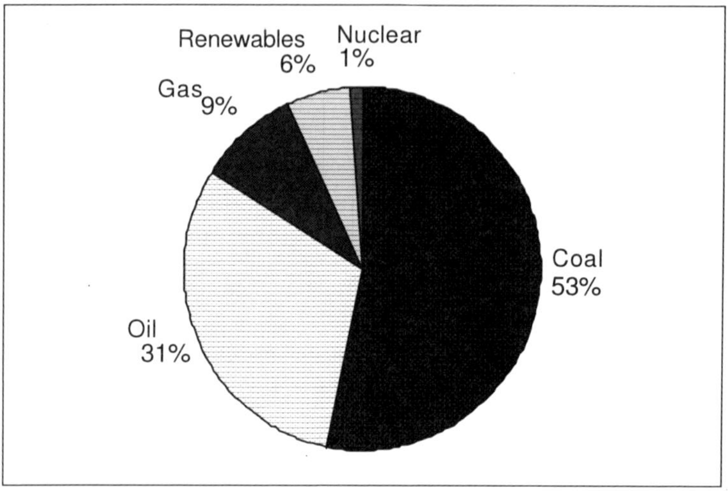

Data Source: Infraline Energy.

Figure 2.2: India's Primary Energy Mix, 2030

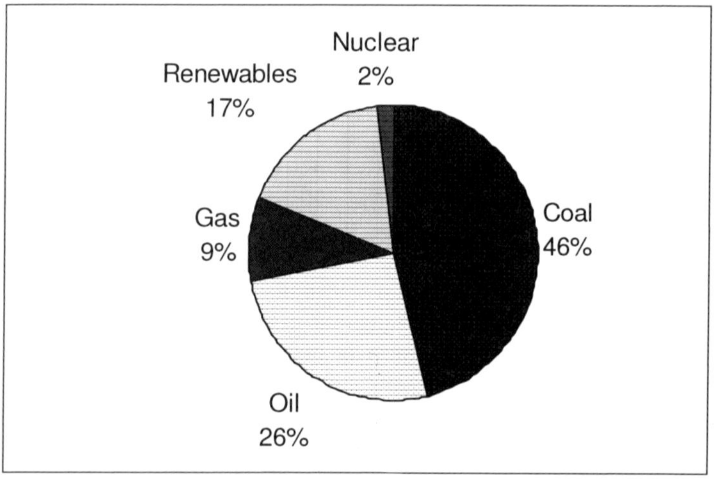

Data Source: Infraline Energy.

The share of renewables is likely to go up.[30] Coal and oil, are expected to remain the dominant fuels in the next couple of decades. India's per capita energy and electricity consumption is woefully low. According to *International Energy Agency's* selected indicators of 2010, India's electricity consumption stood at 644KWh/yr units a year in comparison to the world's average of 2898 KWh/yr. India's current per capita electricity consumption is 778 KWh per year. The developed countries' energy consumption is exceptionally high: OECD 8315 KWh, Australia, 10,063KWh Canada15145 KWh, People's Republic of China 2942KWh, Chinese Taipei 10237 KWh, USA 13361 KWh, U.K. 5741 KWh South Korea 9851 KWH, Japan 8399 KWh, Germany 7217 KWh, Russia 6460 KWh and Brazil 2384 KWh.[31] India's energy demand is expected to grow at the rate of 6.5 per cent per year in the Twelfth Five Year plan. Consequently India's import dependence too would increase.

Table 2.1: Projected Primary Commercial Energy Requirement (Million tonnes of oil equivalent)

	*2010-11**	*2016-17@*
Oil	164.32	204.80
Of which imports	125.5 (76.4%)	164.8 (80.5%)
Natural Gas & LNG	57.99	87.22
Of which imports	10.99 (19%)	24.8 (28.4%)
Coal	272.86	406.78
Of which imports	54 (19.8%)	90 (22.1%)
Lignite	9.52	14.00
Hydro	10.31	14.85
Of which imports	0.48 (4.6%)	0.52 (3.5%)
Nuclear	6.86	9.14
Renewables	0.95	1.29
Total Energy	522.81	738.07
Total Imports	190.97	280.12
% of Total Energy	**36.53**	**37.95**

Source: Government of India, Planning Commission, *Faster, Sustainable and More Inclusive Growth: An Approach to the Twelfth Five Year Plan (2012-17)* October 2011, p.28. http://planningcommission.gov.in/plans/planrel/12appdrft/appraoch_12plan.pdf Accessed on July 27, 2012.

Power Sector

Like other sectors, the power sector too, has evolved from a controlled to competitive market driven regime. Its objective is to provide affordable, reliable and quality power at reasonable price to all sectors of the economy.

Massive additional generation capacity would be needed to meet the rapidly growing need for electricity. This, however, should be done while keeping in mind the overall objectives of sustainable development. The objective is not only to enhance the power generation capacity but also to make it available to the poor sections of the population and to distribute it across all geographical boundaries. By the end of June 30, 2013, the Installed power generation capacity has increased to about 2,25,793.10 MW comprising of Coal based 132,288.39 MW (58.59%), gas based 20,359.85 MW (9.02%), diesel based 1,199.75 MW (0.53%), hydro 39,623.40 MW (17.55%), nuclear 4,780 MW (2.12 %) and Renewable Energy Sources 27,54171 MW (12.20%).[32]

Table 2.2: Total Installed Capacity as on June 30, 2013

Fuel	*MW*	*Percentage*
Total Thermal	153847.99	68.14
Coal	132,288.39	58.59
Gas	20,359.85	9.02
Oil	1,199.75	0.53
Hydro (Renewable)	39,623.40	17.55
Nuclear	4,780.00	2.12
RES** (MNRE)	27,541.71	12.20
Total	**2,25,793.10**	**100.00**

Renewable Energy Sources (RES) include SHP, BG, BP, U&I and Wind Energy
SHP = Small Hydro Project, BG = Biomass Gasifier, BP = Biomass Power,
U&I = Urban & Industrial Waste Power, RES = Renewable Energy Sources
Source: Power Sector at a Glance "ALL INDIA".

Total Installed Power Generation Capacity as on 30.06.2013 and Future Prospects: As on June 30, 2013 the total installed capacity for electricity generation in the country has reached 2, 25,793.10 MW from a mere 16,271 MW in 1971 registering an impressive compound annual growth rate

Figure 2.3: Total Installed Capacity as on June 30, 2013

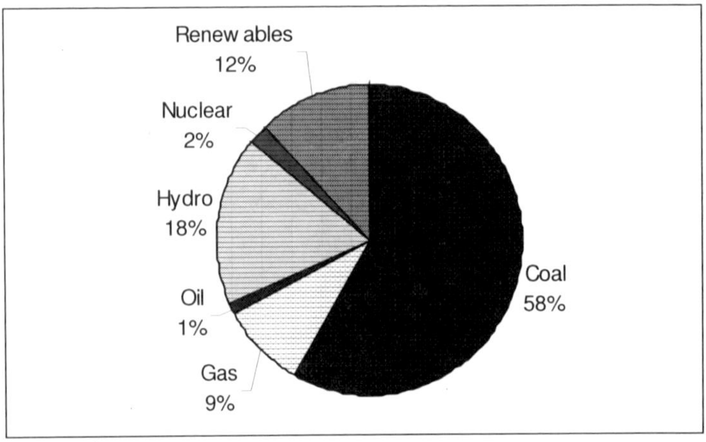

(CAGR) of about 6.4 per cent. The Indian power sector has made substantial progress both in terms of enhancing power generation and in making available power to widely distributed geographical boundaries. The per capita consumption of electricity in the country has increased from 15.6KWh/per year units in 1950 to about 766 KWh/per year units during the year 2009-10. By 2012 it has reached 778 KWh/per year units.

The capacity addition requirement during 12th Plan ending March 2017, according to the Planning Commission's Working Group has been set at 75,785 MW. In accordance with the Low Carbon Growth Strategy of the government of India, priority is to be given to renewable energy sources such as hydro and nuclear generation capacity. Accordingly, a hydro capacity addition of 9,204 MW and nuclear capacity addition of 2,800 MW are considered feasible and thus must be achieved during 12th Plan. Gas based capacity has been set at only 1,086 MW, since gas for these projects is assured as it is available from local sources. Besides this, 1200 MW would be imported from Bhutan. The balance capacity addition of 62,695 MW to meet the demand would be coal based.

The coal requirement for the target has been worked out at 842 (MT). The projected addition of 18,500 MW grid interactive renewables capacity is considered to be quite ambitious. The composition of this renewables according to the Ministry of New and Renewable Energy (MNRE) would be: 11,000 MW wind, 1,600 MW small hydro, 2,100 MW Biomass power, Baggase Cogeneration and waste to energy put together and 3,800 MW Solar.[33]

Fossil Fuels

Coal and Lignite: India has good reserves of coal and lignite. As on March 31, 2011 the estimated reserves of coal were around 286 billion tonnes, an addition of 9 billion over the previous year when they stood at around 277 billion tonnes marking an increase of 3.1 per cent. The estimated reserves of lignite as on March 31, 2011 were 41 billion tonnes, of which 80 per cent were in the southern state of Tamil Nadu. The increase in the estimated reserves of lignite during the year 2010-11 was 2.4 per cent[34] India stands sixth in terms of coal reserves. Russia occupies the top position with estimated 6500 billion tons of coal reserves followed by China, USA, Australia and Canada.

Figure 2.4 Estimated Reserved of Coal as on March 31, 2011

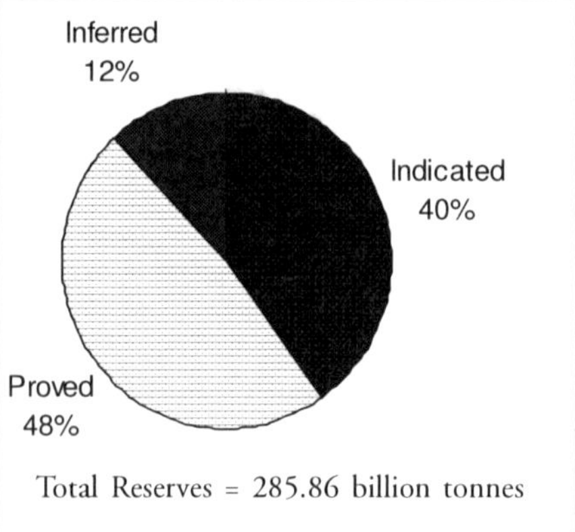

Figure 2.5: Estimated Reserved of Lignite as on March 31, 2011

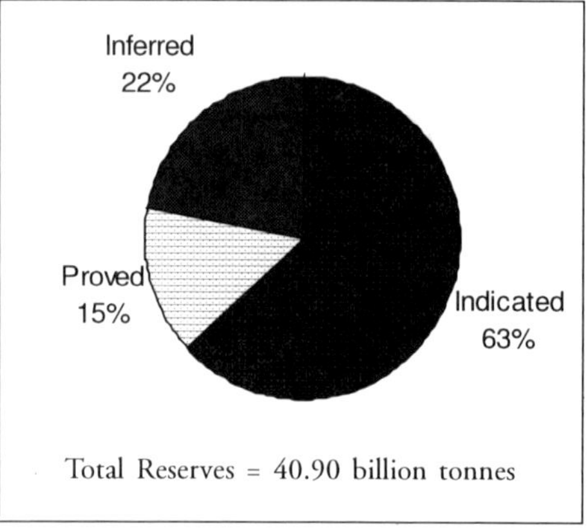

The Planning Commission's Integrated Energy Policy report has estimated that proven reserves of coal at current level of consumption can last for 80 years. In case all the inferred reserves materialise than it can probably last 140 years, at current levels of extraction. However, as coal and lignite consumption is sure to increase in the coming years so the expected life span of the coal reserves would be much less than expected. Thermal power plants using coal account for more than 55 per cent of the power generation capacity. Indian coal has a high content of ash and low calorific value with an average of 4000 Kcal/kg as compared to 6000 kcal/kg in imported coal.

The quality of coal has been deteriorating primarily due to the open cast mines. Consequently, the improvement in productivity is marginal. A number of factors are supposed to have contributed towards hampering the growth of the coal industry.[35] The Coal Mines Nationalisation (Amendment) Bill, 2000 was introduced in the Rajya Sabha on April 24, 2000, but has yet to be passed because of the opposition from the trade unions. The central government is reported to have decided to amend the Coal Mines Nationalisation Bill, 1973. The chances of it being carried are remote as there is still strong opposition from the trade unions.[36] Despite the amendment being pending before the Parliament, the government has formulated the mining policy within the bounds of the existing law and several coal blocks have been offered to private entrepreneurs to exploit coal for their own consumption. FDI in coal mining has been allowed and mining by joint ventures has been permitted. Coal blocks have also been allocated to other PSUs under central and state governments for coal mining.[37] The mismanagement of coal block allocations has hampered the coal mining and the cases of corruption in coal block allocations are under investigation.

Import of Coal: In order to bridge the gap between demand and coal availability, power utilities are expected to import around 159 MT of coal to meet shortages in coal supply from CIL. However, the blending of such huge quantities of imported coal may not be feasible as only a maximum 15 per cent blending of imported coal is possible in existing boilers. This quantity would be in addition to 54 MT coal likely to be imported by thermal power Stations designed for imported coal. Therefore, the total quantity of coal, expected to be imported, is about 213 MT by 2017 when the 12th plan ends.[38] The efficiency of energy conversion and the limiting of carbon emissions can be achieved to a certain extent by developing clean coal technologies. Some of these technologies such as coal bed methane[39] and in-situ gasification[40] can be helpful in developing additional sources of energy by exploiting the un-mineable and deep-seated coal reserves and liquefaction of coal. However, these technologies are still in the process of development.[41] The syn-gas produced by Underground Coal Gasification (UCG) can be used to generate electricity through combined cycle. It can also be used chemically to produce synthetic natural gas (e.g. Great Plains Gasification Plant in North Dakota). It may also serve as a feedstock for methanol, gasoline, or diesel fuel production and even as a hydrogen supply. Currently, this technology could be deployed in the highly populated areas of both eastern and western India, thus reducing the overall energy demand

Exploration and Production of Hydrocarbons (Oil and Gas): The estimated reserves of crude oil and natural gas in India as on March 31, 2011 stood at 757 million tonnes (MT) and 1241 billion cubic metres (BCM), respectively.[42] Gas is expected to play an important role in the country's energy mix. India is expected to capitalise on indigenous gas reserves that are considerably larger than India's oil endowment. There is scepticism regarding the prospects of development of shale gas in India.[43] However, large off shore gas finds have helped boost India's reserves from 0.8trn cu metres (28.25trn cu ft) in 2004 to 1.2trn cu metres (42.37trn cu ft) in 2011. As per the BP Statistical Review of World Energy (June 2012) production is expected to increase from an estimated 46.4mtoe in 2011 to 57.4mtoe in 2020, provided new discoveries are made. However, more gas imports would be required for energy security. Liquefied Natural Gas (LNG) will be one option and Australia could be an important source. But India will have to face competition from its Chinese counterparts to secure LNG contracts which come with high price tags. The Turkmenistan-Afghanistan-Pakistan-India (TAPI) pipeline would be an important development but the prospects of its materialisation at present seem to be very remote.[44]

Figure 2.6: Estimated Reserve of Crude Oil as on March 31, 2011

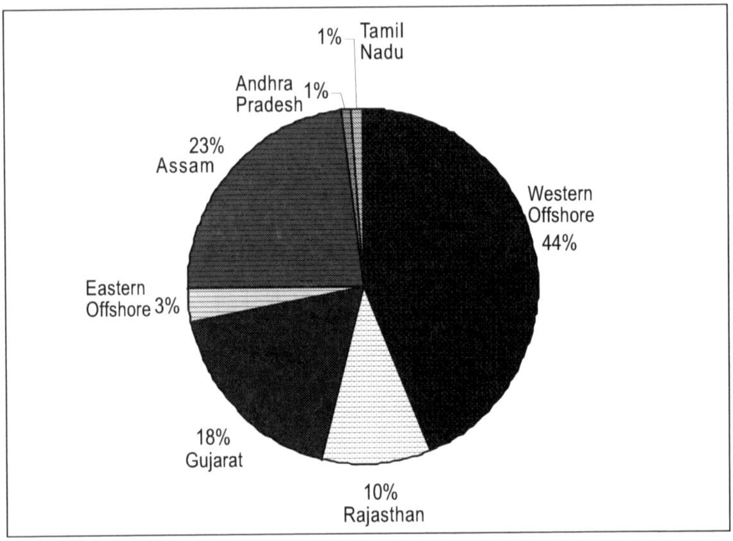

Total Reserves = 757.44 Million Tonnes

Figure 2.7: Estimated Reserve of Natural Gas as on March 31, 2011

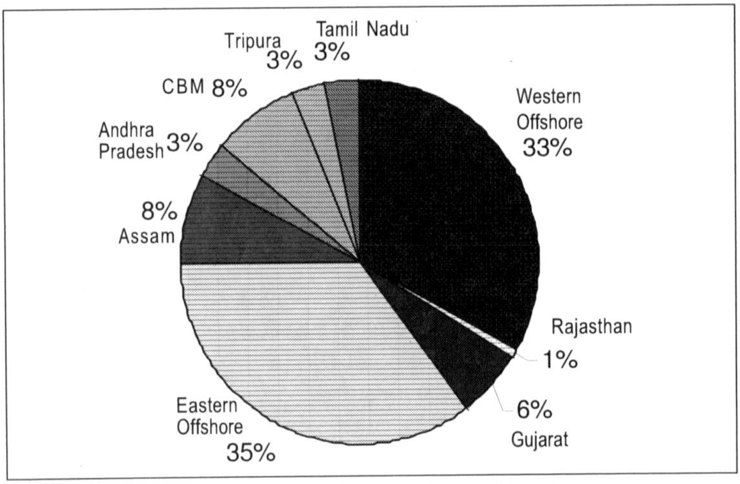

Total Reserves = 1240.92 Billion Cubic Metres

Non-Fossil Fuels: New and Renewable Energy Sources

Renewable Energy: There is great potential for the generation of renewable energy from various sources—wind, solar, biomass, small hydro and cogeneration bagasse. The total potential for renewable power generation in the country as on March 31, 2011 was estimated at 89,760 MW. This includes an estimated wind power potential of 49,132 MW (55%), SHP (small-hydro power) potential of 15,385 MW (17%), biomass power potential of 17,538 MW (20%) and 5000 MW (6%) from bagasse-based cogeneration in sugar mills. The geographic distribution of the estimated potential across states reveals that Gujarat has the highest share of about 14 per cent (12,489 MW), followed by Karnataka with 12 per cent(11,071 MW) and Maharashtra with 11 per cent (9596 MW), mainly on account of wind power potential.[45]

There has been a steady growth in renewables—both incremental and cumulative—between 2003-12 and has the further potential to grow at 14 per cent in the next decade. The total installed capacity of grid interactive renewable power, which was 16,817 MW as on March 31, 2010 has gone up to 26,677.10 MW as on December 31, 2012. The share of wind, solar, small hydro and biomass was 18,420 MW (69.04%), 1,176.25 MW (4.40%), 3,496.14 MW (13.10%), and 3,584.31 MW (13.43%) respectively.[46] The current figures show that the fastest growth has been in wind energy followed by small hydro and biomass. Solar energy has shown the least growth but has the potential to grow in future. The geo-thermal potential too has been identified at around 10GW but progress has been slow.

The Ministry of New and Renewable Energy had set the following targets for the grid interactive renewables for the 12th Five Year Plan (2012-17): Wind power 11.000 (MW); Biomass; Baggasse Co-generation, Biomass Gasifiers 2,100 (MW); Small Hydro (up to 25 MW) 1,600 (MW); and Solar Power 3,800 (MW)—total of 18,500 MW. The growth of the renewable energy has been linked with the National Action Plan for Climate Change which calls for the growth in sustainable renewables through 2020. The government has taken important steps to promote the growth of the renewable energy by giving subsidies in the form of (1) Renewable Energy Certificates (RECs)—a market based instrument to promote renewable energy and facilitate Renewable Purchase Obligations (RPO); and the Feed-in-Tariff mechanism under which a preferential tariff is fixed by the state for the procurement of power from renewable energy sources.[47]

The National Action Plan for Climate Change (NAPCC) had set a target of 5 per cent renewable energy purchase for the FY 2009-10 with a view to increase the share of renewables in the total energy consumption. This has been further extended. It further envisages that this target will raise by one per cent per annum for the next 10 years. The NAPCC, thereby, envisages renewable energy to contribute approximately 15 per cent to the energy mix and to achieve this target it mandates an instrument called the renewable energy certificates.

The policies under Electricity Act 2003, as well as NAPCC provided a roadmap for increasing the share of renewables in the total generation capacity of the country. But the Renewable Energy (RE) resources are not evenly spread across the different parts of the country, such as Delhi, where there is hardly any potential for RE. Therefore State Electricity Regulatory Commissions in such states are inhibited from specifying higher Renewable Purchase Obligations (RPO). It may be mentioned that Section 86 (1) (e) of the Electricity Act 2003 provides for cogeneration and generation of electricity from RE by providing "suitable measures for connectivity to grid", "sale of electricity to any person" specifying the quota obligation for consumption of RE (Renewable Purchase Specification (RPSP) or Renewable Purchase Obligation (RPO).

Renewable generators have two options with regard to the electricity generated, by them—either to sell the renewable energy at preferential tariff fixed by the concerned Electricity Regulatory Commission; or to sell the electricity and the environmental attributes incorporated in it as renewable energy. In case of the second option the environmental attributes can be exchanged in the form of Renewable Energy Certificates to be issued by the CERC to the renewable generators. One REC would be equivalent to 1 MWh of electricity injected into the grid from the renewable source.

This REC can be traded or exchanged only in Power Exchanges approved by the CERC. There are two price bands for exchange: one is the floor price which is the minimum price; and the second is the forbearance (ceiling) price or the maximum price, that is determined by the CERC from time to time. The distribution companies, Open Access Consumers, Captive Power Plants (CPPs) can buy these RECs to meet their Renewable Purchase Obligations (RPO).[48] RECs can be used by the obligated entities to demonstrate compliance with regulatory requirements, such as Renewable

Purchase Obligations. There are two categories of certificates: Solar certificates issued to eligible entities for generation of electricity based on solar as renewable energy source; and Non-solar certificates issued to eligible entities for generation of electricity based on renewable energy sources other than solar. The RECs have a validity of 365 days after the date of issue. And the participation is voluntary. They can be traded only at the Power Exchange and cannot be banked or borrowed.[49] The cost of electricity generated from renewable energy sources has two components, one, its cost equivalent to conventional energy sources and two, cost of environmental attributes. The environmental attributes can be exchanged in the form of Renewable Energy Certificates (REC).[50]

Energy Generation

The Indian government has taken concerted steps to promote the development of renewable energy through tax incentives and purchase of electricity generated through renewable. The Electricity Act 2003 envisages purchase by power distribution utilities of a certain percentage of electricity generated through renewable energy sources. The REC and the Indian Electricity Grid Code-2010 (IEGC) are directed towards the promotion of renewable energy in India and for achieving the targets set by respective SERCs for renewable purchase obligations. This is also meant to help minimize cost of power procurement and leading to efficient resource utilization across the country and provide incentive for investment in appropriate technologies.[51]

A major drawback of the REC trading market has been that transactions have been mostly carried out by the government entities, and the corporate houses have not been keen on this. However, it is suggested that more corporate houses should volunteer to trade in the REC market which would help expand the REC market in India. For the success of REC mechanism, it is imperative that public sector units should be encouraged to get involved and for this more incentives should be provided to the buyers.[52] The public sector units across the country should also buy REC/green energy and make it a part of their corporate social responsibility. Moreover, incentives should be given to the corporate houses, to motivate them, to get involved in the REC. The prospects of financial benefits in the form of tax rebates should be explored.[53]

The Central Electricity Regulatory Commission (CERC) determines the floor price and the forbearance price for dealing in certificates. The CERC

Figure 2.8: REC Processes

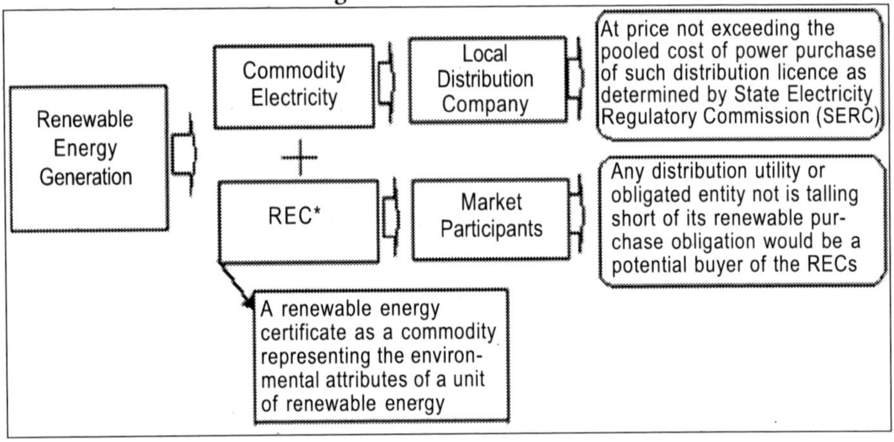

Figure 2.9: REC Processes

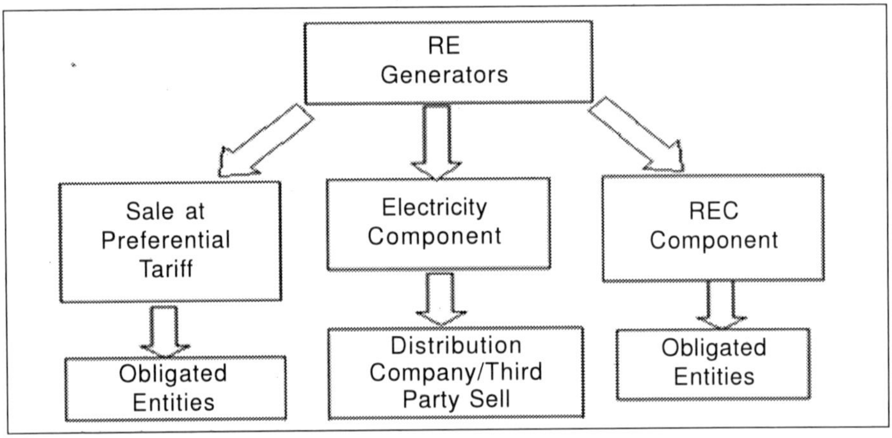

has declared the new floor and forbearance price for REC, Solar and Non-Solar for the control period of 2012-17. The new floor and forbearance prices are: Forbearance price for non-solar REC, Rs 3300 per MWh; for solar it is Rs. 13,400 per MWh; the floor price for non-solar REC is Rs. 1,500 per MWh while for solar it is Rs. 9,300 per MWh.[54] The earlier REC prices were valid up to FY 2012, which were as: Forbearance price for non-solar REC was Rs. 3,900 per MWh; for solar, Rs. 17,000 per MWh. The floor price for non-solar REC was Rs. 1,500 per MWh and Rs. 12,000 for solar.

The reluctance to buy REC by state "obligated entities" is attributed to the non existence of any liability or penalty for the non-fulfilment of the RPO. However, it is argued that some of these constraints have been addressed as renewable generators are free to sell power and they also have an additional revenue stream from the sale of REC's. Renewable energy faces multiple challenges from transmission to bank-ability. Some of these are: (1) the problem of Grid connectivity; (2) Difficulties of Financing; (3) Market uncertainties; (4) Resource availability; and (5) problems of access to technological developments

Off Grid Captive Power installed renewable Capacity: An estimated 52 per cent of the rural households in India are electrified but the percentage of off-grid households in India stands at at staggering 94 per cent. Off-grid installed renewable energy capacity as on December 31, 2012 was as follows: Waste to energy urban/rural 113.60 (MWeq); Biomass (non-bagasse) cogeneration 426.04 (MWeq); Biomass-Gasifier-rural/16.696 (MWeq); Biomass-Gasifiers-Industrial138.90 (MWeq); Aero-Generators/Hybrid system1.74 (MWeq); SPV>1kW106.33 (MWeq) Watermills/micro Hydel 2121 Nos; Total—803.306 (MWeq).[55] The development of the off-grid energy system is equally important as it is a mechanism by which rural household can at least access power for electricity and other household chores, but this too is facing innumerable problems.[56]

Use of Renewable Energy for Telecom Towers: The off-grid renewable energy can also be used for powering the telecom towers particularly in rural areas where most of the telecom towers have poor grid connectivity. So far most of the telecom towers are dependent upon diesel generated power and diesel consumption is set to increase with the growth of the telecom sector.[57] The Biomass Power Programme has been implemented with the objective of harnessing grid quality power from biomass resources through various conversion technologies. Biomass materials used for power generation include bagasse, rice husk, straw, cotton stalk, coconut shells, soya husk, de-oiled cakes, coffee waste, jute wastes, groundnut shells, and saw dust etc.

Small Hydro Power Projects (SHP): The responsibility for developing small hydro power projects is vested with India's MNRE. Hydro power projects up to a capacity of 25 MW are put in the category of SHP. The estimated potential for SHP/Mini projects is about 15,380 MW. The total installed capacity of the Small Hydro projects as of December 31, 2012 stood at 3496 MW.

Wind Energy: India and has become the fifth largest wind power producer in the world, after USA, Germany, Spain and China. As of December 31, 2012 a total capacity of 18,420 MW has been installed. This constitutes more than 69 per cent of India's overall renewable energy and 8.5 per cent of the total installed power generation capacity in the country. Beginning in the early 1990's, there has been remarkable growth in wind energy which has now become the largest contributor of renewable generation in the country. It is a renewable energy technology which generates grid connected power. This is an impressive performance in terms of its global standing. India's seashore wind power potential has been estimated at 49,130 MW. This would require about 2 per cent land of all states except the Himalayan states, the North-eastern states and Andaman Nicobar. For these states only 0.5 per cent will be required.[58] It is expected that about 7,500 MW wind power capacity will be installed in India between 2012 and 2015, taking the total installed capacity beyond 22 GW.[59]

According to Indian Wind Turbine Manufacturers Association (IWTMA) India has the potential to develop 65 to 70 GW wind power. The MNRE has estimated it at about 50 GW, while the World Institute for Sustainable Energy (WISE) has estimated that with larger turbines and greater land availability and expanded resource exploitation the potential could go up to as high as 100 GW. This could be a significant contribution to the country's overall electricity generation. However, since most of the high wind density sites have already been exploited, the only available option would be the repowering of existing wind farms by upgrading them with better design as well as more efficient turbines that would result in the wind farms achieving a higher plant load factor (PLF) and consequently higher realisation of revenue. MNRE is also exploring the possibilities of off shore wind farms to boost wind energy capacity.[60] The growth of the wind energy has been possible because of the mechanism of Feed in Tariff,[61] the subsidies in the form of Accelerated Depreciation[62] and Generation Based Incentives[63]. Consequently tariff rates in different Indian states for 1 KWh of electricity are: Andhra Pradesh Rs 3.50; Gujarat Rs 4.23; Karnataka Rs 3.70; Kerala Rs 3.60; Madhya Pradesh Rs 4.35; Punjab Rs.5.96; Rajasthan Rs.5.18; Tamil Nadu Rs.3.51; and West Bengal Rs.4.87.[64]

Solar Energy: As of December 31, 2012 solar power projects of around 1176.25 MW capacities have been commissioned in the country. These include 153 MW grid solar power projects under, Jawaharlal Nehru National Solar Mission (JNNSM), and the balance have been set up under the states' grid-connected solar programmes. The JNNSM was formally launched on January 12, 2010 by Prime Minister Manmohan Singh who called for the creation of "solar valleys" in India on the lines of Silicon Valley, to contribute to national, as well as global efforts, towards combating climate change. The mission has been planned to create an installed capacity of 20,000MW (20GW) by the end of 13th Plan.[65] The NSM is divided into three phases, and aims to meet targets of 20,000 MW of grid connected solar power and 2,000MW of off-grid installations by 2022. But it remains uncertain whether the projected allocated under NSM will be operational within the projected timeline. Phase 2 targets an ambitious minimum addition of 3,000 MW and with the potential of reaching 10,000 MW by 2017.[66] However, there are still many hurdles in the way of solar energy and there is need for concerted efforts on the part of state and central government to promote solar energy production. As the Mission moves into Phase 2, it is expected that the ministries and the government agencies and stakeholders will need to:

> ...develop effective solutions for more credible project bids, enforceable Renewable Purchase Obligations (RPOs) and Renewable Energy Certificates (REC), strong financial structures, increased domestic manufacturing, and reliable power evacuation and transmission.[67]

Off-grid solar energy which so far has not received due attention should also become the focus of investment. However, the problems being faced by the solar power industry need to be sorted out. These are: Easy access to banking finance or bank-ability of solar projects;[68] reliable power evacuation and transmission;[69] issues relating to land acquisition; community involvement and habitat protection; greater flow of information through benchmark transparency and monitoring the mission progress; and strategic financing.

Table 2.3: National Solar Mission Targets 2010 to 2022

Solar Technology	**Phase 1 (2010-13)**	**Phase 2 (2013-17)**	**Phase 3 (2017-22)**
Grid-connected/ rooftop	1,000 MW – 2,000 MW	4,000 MW – 10,000 MW	20,000 MW
Off-grid solar applications	200 MW	1,000 MW	2,000 MW
Solar hot water collectors	7 million sq. meters	15 million sq. meters	20 million sq. meters
Rural solar lanterns/ Lighting	N/A	N/A	20 million systems

Source: Jawaharlal Nehru National Solar Mission: Towards Building Solar India.

The price of photovoltaic (PV) panels has come down drastically and grid power, particularly the thermal power, is becoming costly. Solar power is becoming cost effective and has also reduced the timeline to achieve grid parity. The importance of the concentrated solar power (CSP)—a technology that produces electricity by concentrating solar energy on a single focal point —also cannot be ignored. This concentrated energy is used to heat water and generate steam which in turn is used to run the turbines that produce electricity. Different techniques can be used to achieve concentration of solar power.[70] However, it is not PV versus CSP. India needs both PV and CSP as it needs affordable electricity as well as the stability of its grid infrastructure. The CSP is also facing problems, such as the non availability of non-recourse financing [71] as CSP projects and CSP developers can get finance only if the parent company furnishes corporate guarantee which is affected by the poor quality of Direct Normal Irradiation (DNI)[72] data. This reduces the confidence of the investors. Direct Normal Irradiance (DNI) is the amount of solar radiation received per unit area by a surface that is always perpendicular (or normal) to the sun's rays that fall in a straight line from the direction of the sun at its current position in the sky. The unreliability of DNI data is further affected by the existence of aerosols.[73] Apart from the poor DNI data, the lack of sufficient ground level data on aerosols (including dust particles, sand, and other airborne particles) further reduce the reliability of the available DNI data; a strong REC market is essential for bolstering investment in renewable energy in general, and solar energy in particular; urgent need to strictly enforce RPOs to meet the NSM targets particularly in the second phase when the target is 6000 MW. In addition, states should take strict initiatives to enforce RPO. In order to enforce compliance even the state's central funding could be linked with the meeting of RPO obligations. Public utilities and state electricity distribution companies should be pressurised to meet RPO targets thereby creating demand for RECs.[74]

Nuclear Power

Despite the accident in Fukushima, and the understandable caution worldwide on nuclear energy, it will remain important for India as non-polluting source energy and as the best alternative for environmental friendly economic development and growth, not contributing to the climate change. India has an ambitious nuclear energy programme, which started with Dr Homi Bhabha, and is progressing rapidly. It has attained self-reliance in all aspects of the close fuel cycle. The Nuclear Suppliers Group's waiver for India in 2008 gave a new impetus to its nuclear energy expansion, opening the gates for its participation in international nuclear commerce. A large number of bilateral agreements have been signed in last few years. A positive aspect of this development is that the nuclear energy sector will help to create a high technology nuclear industry which will provide large number of quality jobs to Indian youth and enable the export of nuclear technology. However, there is an element of scepticism about the growth of the indigenous nuclear industry as it is beset with a number of problems.

Out of India's total installed power generation capacity of 2,25,541.71MW as on June 30, 2013, the share of nuclear power was 4,780.00 MW, which with the linking of Kudankulam-1,VVER-1000, has gone up to 5,780MW, which is less than 3 per cent.[75] Currently there are 21 nuclear reactors[76] in operation in India. India has an ambitious plan to raise the nuclear power generation capacity to 20 GW (e) by 2020, 63 GW (e) by 2032 and 275 GWe by 2052 and increase nuclear energy's present share to about 25 per cent by 2052. Seven reactors of 5300 MW capacity are under construction.[77] According World Nuclear Association, 18 reactors are planned for the next phase of expansion[78] and about 39 more nuclear energy parks are proposed.[79] The government of India has revised the nuclear energy targets as follows: the installed capacity of 10,080MW will be achieved by 2017 and 27,480 MW will be achieved by 2023-24.[80]

During the Twelfth Plan (2012-17) NPCIL has planned to initiate the development of several new nuclear power projects. These are: Eight indigenous PHWRs of 700 MW each—Gorakhpur 1 and 2 (Haryana); Chutka 1 and 2 (Madhya Pradesh); Mahi Banswara 1 and 2 (Rajasthan); and Kaiga 5 and 6 (Karnataka).Eight light water reactors each of 1000MW or more with international cooperation—Kudankulum 3 and 4 (Tamil Nadu); Kovvada 1 and 2 (Andhra Pradesh); Jaitapur 1 and 2 (Maharashtra); and Chhaya Mithi Virdhi 1 and 2 (Gujarat). Two fast breeder reactors of 500 MW each—Kalpakam 1 and 2 (Tamil Naidu); and one advanced heavy water reactor of 300 MW.[81] Apart from importing nuclear reactors with international cooperation, NPCIL is to develop nuclear reactors through joint ventures with Indian companies. It has already signed agreements for forming joint venture companies with the National Thermal Power Corporation Limited, the Indian Oil Corporation Limited and National Aluminium Company of India Limited. The DAE is also working closely with the nuclear industry to develop an indigenous manufacturing base for nuclear equipment and components.[82]

However, the development of planned capacity is facing several challenges such as: issue of land acquisition; public non-acceptance; and inadequate industry capacity and capability. Of late the lack of public acceptance and local opposition to nuclear power plants has emerged as major stumbling blocs to the achievement of targeted goals. The table below shows the projected nuclear power capacity addition (MW).

Figure 2.10: Projected Nuclear Power Capacity Addition (MW)

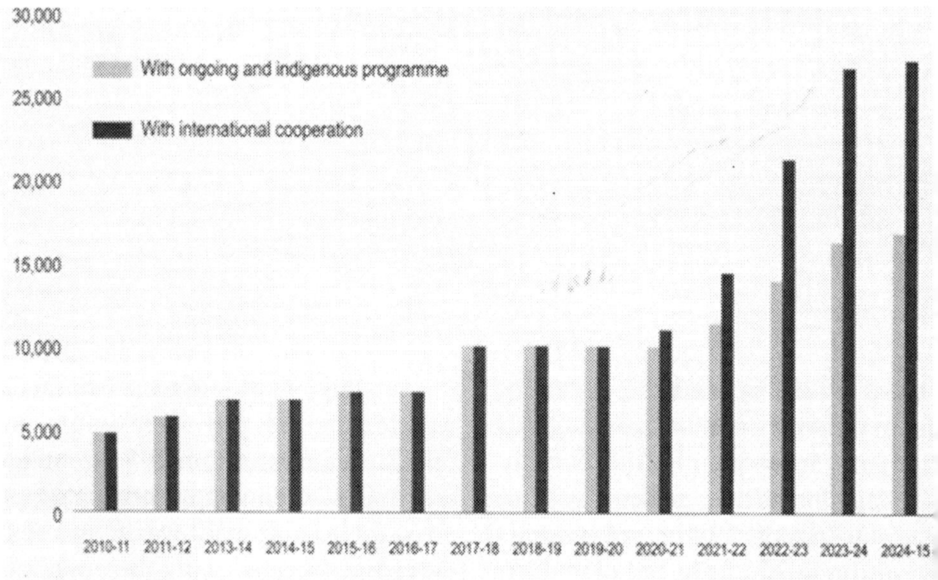

Source: R.B. Grover and N. Nagaich, "Nuclear Plans", Power Line, January 2012.

Two VVER-1000 power reactors at Kudankulam—after facing strong protests from the local population—are ready for commissioning. Six (1,600x6 MW) European Pressurised Reactor (EPR) design reactors that are being imported from France and are to be located at the Jaitapur nuclear Power Station are also the targets of anti-nuclear protests. Another 20 nuclear reactors are already being planned. There are proposals for 40 more reactors in the long run.

The high targets set by the government are sought to be achieved by indigenous technological development and by pursuing a three stage nuclear development plan and international nuclear cooperation, including the import of nuclear reactors. India's three-stage nuclear development programme, as such, is soundly structured. However, its success to a great extent would depend on the crucial technological breakthroughs in India's indigenous nuclear energy programme. India's technological advancement in the nuclear arena would determine its status as an advance nuclear supplier state from that of as nuclear recipient state.

India's nuclear programme by all standards is a bold initiative. It is bold because other countries have not experimented with it, so to that extent, we should be proud of our scientists. Scepticism is expressed about India's breeder programme, because it has very little international experience. The reluctance to pursue a breeder programme is because it is a proliferation prone technology but India has the confidence to pursue it. As regards Thorium, India would be pioneering this fuel cycle so the question of international collaboration in this case—at least at present—does not arise as there is hardly any country with any experience in the Thorium fuel cycle technology.

In the case of breeder reactors, doubts are expressed with regard to the fuel it generates compared to what it uses. It is accepted that in a breeder reactor, the "conversion ratio" or "breeder ratio" is higher than one. The "breakeven" occurs when the "conversion ratio" is exactly one, and the reactor produces exactly as much fissile material as it uses. In commonly used Light Water Reactors the conversion ratio is approximately 0.6, while in Pressurised Heavy Water Reactors (PHWR) running on natural uranium the conversion ratio is 0.8.

The reluctance on the part of other countries to collaborate with India's nuclear programme, despite it being innovative and unparalleled, can be because of non-proliferation concerns. However, the Indo-US civil nuclear deal of 2005 and the subsequent NSG waiver in October 2008 unconditionally lifted the restrictions on nuclear trade with India bringing to an end the 34 year sanctions on nuclear commerce with India. Since the removal of the sanctions the Department of Atomic Energy (DAE) and other entities such as NPCIL, are making renewed efforts to formulate an impressive plan for the development of nuclear energy in India. They are also taking measures to inject transparency into the nuclear programme so as to enhance mutual trust within the country and outside, which could lead to collaboration with other countries.

It is also contended that the indigenous nuclear industry has not been able to resolve several technological problems, and it is hampered because of lackadaisical growth and development. However, a reality check of the indigenous nuclear industry would attribute this flaw to the lack of demand, as the Indian nuclear industry withered due to technological difficulties, non-proliferation constrains and the insufficient fuel supplies. The situation has changed after the civil nuclear deal of 2005 and NSG waiver. The government's policy measures encouraging public–private participation in the nuclear engineering sector would give the desired boost to the nascent private nuclear engineering industry. Some private players such as Larsen and Toubro (L&T), Walchand Industries Limited (WIL), Godrej and Avsarala Technology Ltd. are already providing the core components for the nuclear reactors. The further advancement of the nuclear engineering and nuclear reactor manufacturing sector would enhance India's international profile as a supplier of nuclear energy design, construction, operation, safety and other nuclear related services. This would also help India to position itself as an exporter of small and safe nuclear reactors for countries with smaller needs of clean energy. The DAE has plans for importing reactors from various advanced countries in addition to indigenous construction of new nuclear power plants. So far India has signed bilateral nuclear cooperation agreements with nine countries. India is in touch with uranium rich countries like Kazakhstan, Australia and Niger for the import of nuclear fuel.

To achieve these high targets, the DAE would have to identify the key constraints that pose an obstacle to the further expansion of nuclear energy—particularly opposition from the civil society, the technological hitches, resources, fuel supplies, safety and security of nuclear reactors, waste management and security of the radioactive waste, legal issues, such nuclear liability, and preparedness to meet the nuclear disaster in case of crisis. All these concerns have made the sceptics raise doubts about the feasibility and achievability of these targets. Other doubting voices have raised concerns about the safety of India's reactors.

A Comparison of Different Sources of Energy

After the Fukushima nuclear accident questions were raised about nuclear energy's desirability as a safe and pollution free energy option. But at the same time it is pitched as non-desirable and a dangerous source of energy because of the issues of radiation and radioactive waste management. All over the world, the issue of energy choices has taken the centre stage and the contestation has become much sharper. Consequently, comparisons are being made between the various sources of energy in terms of environmental effects, consistency, its cost and availability in bulk. No doubt nuclear energy is the most desirable when compared with fossil fuels (coal, oil and gas). However, the renewable sources of energy particularly solar and wind have emerged as the preferred sources of energy. Anti-nuclear groups all over the world, particularly the European countries, are making a strong case for renewables and are denigrating nuclear energy as the most dangerous source of energy. However, from the point of view of energy deficient countries like India, it is not a question of renewable or nuclear, rather it is both renewable and nuclear, as both would contribute towards achieving the goal of energy security.

Nuclear vs. Fossil fuels

A comparative analysis of Green House Gas (GHGs) emissions from various fuels shows that the worst pollution comes from the direct burning of coal as one KWh of electricity generated from coal produces about 790 to 1017 grams of CO_2. Indirect burning produces between 178 to 289 grams of CO_2 per kWh of electricity. Direct burning of gas produces GHGs i.e. 362 to 575gms of CO_2 per KWh of electricity produced. In indirect burning it ranges from 77 to 113 grams of CO_2 per KWh of electricity produced.[83] In reality there is no 100 per cent safe way of generating electricity and no one supporting nuclear power argues that it is 100 per cent safe. However nuclear power is much safer than fossil fuels (coal, oil and gas) which at present are the only options for generating massive amounts of electricity. One important reason for this is that the amount of fuel required to run a nuclear power plant is much less than what is needed for a fossil fuel run plant. For instance to run a 1000 MW coal based power plant for one year requires about 38,000 railroad cars of coal; while in case of an oil fuelled plant 40,000 barrel of oil per day would be required. While nuclear power plant of the equal size would require only 6 truckloads of fuel. It means there would be less pollution on account of transportation of the nuclear fuel, fewer "mining disasters, fewer agonising deaths from ailments such as black lungs, and fewer deaths and injuries in railway accidents while transporting the fuel".[84] It is also easier to protect and safeguard, as nuclear fuel is much more concentrated.

Fossil fuelled power plants release a lot of pollutants into the air. A 1000MW coal based, power plant produces almost 36,500 truckloads of ash containing deadly items as arsenic, lead and mercury. At the same time, the amount of waste generated by a nuclear power plant is very small.[85] Nuclear power has high energy density as each kilogram of uranium fuel contains massive energy, which is almost million times the energy density of fossil fuels. For instance a 1,000MW nuclear reactor would require only 200 tons of milled uranium per year. This makes the fuel cost of generating electricity by using uranium, very low, i.e. only two to four per cent of the final cost of electricity. This low fuel cost and high energy density enables a country to stockpile fuel for a couple of years in a space no larger than a small warehouse, making it highly cost competitive.[86] Equally, high energy density produces relatively very small amounts of waste. The radioactive waste from power plants can be disposed off in geologic repositories at low cost. In case of coal and gas the sheer massive quantities of waste makes it impossible to dispose of all waste. Moreover, it is easier to monitor the environmental impact of nuclear energy than that of other energy sources. The radioactive contamination can also be detected quickly and therefore can be avoided. It is precisely for this reason, it is argued, that there have been no fatalities due to radiation in the Japanese accident.[87]

Least carbon emission energy source: One of the most important factors favouring nuclear power is its low carbon emission. Nuclear energy is the only low carbon electricity source which can be generated and made available 24/7 on a mass scale. Thus, it emerges as the most favoured solution to meet the challenges of the climate change because it lowers carbon emissions in the power sector. Globally nuclear energy produces nearly half of all low carbon electricity and it provides almost 70 per cent of US electricity, generated from low carbon electricity.[88]

Table 2.4: Summary of Lifecycle GHG Emission Intensity[89]

Technology	Mean	Low tonnes CO_2e/GWh	High
Lignite	1054	790	1372
Coal	888	756	1310
Oil	733	547	935
Natural Gas	499	362	891
Solar PV	85	13	731

Technology	Mean	Low tonnes CO_2e/GWh	High
Biomass	45	10	101
Nuclear	29	2	130
Hydroelectric	26	2	273
Wind	26	6	124

It is estimated that each year the US prevents 650 million metric tons of carbon dioxide from being released into the air. It has in operation more than 100 nuclear power plants in operation. At the same time Germany's decision to shut down all its nuclear power plants would result in the release of additional 400 million tons of carbon dioxide by 2020 as per the Breakthrough Institute's analysis.[90] Greenhouse gas emissions of nuclear power plants are among the lowest of any electricity generation method and on a lifecycle basis are comparable to wind, hydro-electricity and biomass. Lifecycle emissions of natural gas generation are 15 times greater than nuclear. Lifecycle emissions of coal generation are 30 times greater than nuclear. There is strong agreement in published studies on life cycle GHG intensities for each generation method. However, the data demonstrates the sensitivity of life cycle analysis to assumptions for each electricity generation source. The range of results is influenced by the primary assumptions made in the lifecycle

Figure 2.11: Lifecycle GHG Emissions Intensity of Electricity Generation Methods

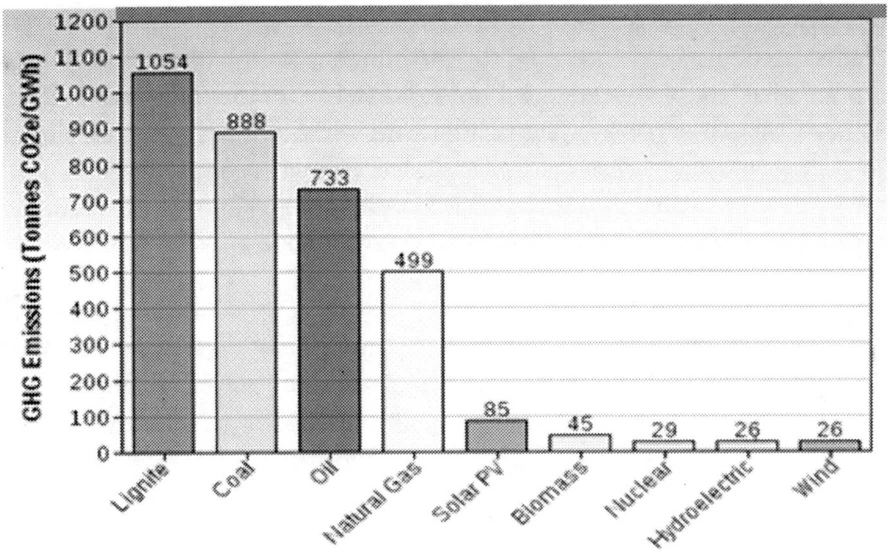

Source: WNA Report.

analysis. For instance, assuming either gaseous diffusion or gas centrifuge enrichment has a bearing on the life cycle result for nuclear.[91]

Low cost electricity from nuclear energy: Another most important factor in favour of nuclear power is the cost of electricity generation. After the initial capital cost involved in the building of the power plant, electricity can be produced at a fraction of the cost as compared to other sources, both fossil and renewable. In the US the nuclear energy facilities are producing electricity at the lowest cost of 2.14 cents per kilowatt hour of electricity. With the imposition of the carbon tax the cost of fossil fuel generated electricity is likely to go further up. Thus, nuclear energy is considered a valuable engine for economic growth apart from reducing air pollution.[92]

It is, however, asserted that nuclear energy is already much more heavily subsidised than the renewable energy. Its real cost is supposed to be much higher than that of the renewables. One reason for this could be that governments seek military–nuclear technology advantages from civilian technology. It is estimated that nuclear power is subsidised between 50 to 90 per cent.[93]

Continuous Safety Evaluation and Improvement

No other source of energy is subjected to such thorough and ongoing safety evaluation and improvement other than the nuclear power. Nuclear facilities are designed with the highest safety standards and after Fukushima extra efforts are being made to strengthen the safety standards of nuclear power plants, particularly to withstand the earthquakes of high magnitude. There is no doubt that nuclear energy is not risk free but for that matter no energy source is risk free. It is argued, that a world without nuclear energy would be fraught with far greater risks of: climate change; economic stagnation; unmet energy needs; dependence on foreign energy and threats to energy security.[94]

Nuclear vs. Renewable: Renewable sources of energy, so far, are on the margins and are likely to continue as their share is not likely to go beyond 5 to 6 per cent even if the 100,000MW is added to India's power mix in the next five decades. The present share hovers around 12-13 per cent. Moreover, when compared with wind, solar, biomass, geothermal, and small hydro, in terms of carbon footprints, nuclear power comes out more or less at par with them. However, in terms of the land required, nuclear has advantage over other green energy sources. While, a 1000 MW nuclear power plant requires one square mile of space, the solar, wind and biomass require 50, 250 and 2,600 square miles of space respectively for an equivalent capacity power plant.[95] Another writer has estimated that a 1,800MW nuclear power plant would require 1.7 square miles of land as compared to 169 square miles for wind and 21 square miles for solar photovoltaic.[96]

Green House Gas emissions: Nuclear and Renewable: Nuclear energy emits less carbon dioxide than the renewable sources of energy. In case of hydro it ranges between 4 to 236gms, solar PV ranges between 100 to 280 gms, wind between 10 to 48 gms while in the case of nuclear it is between 9 to 21 gms CO2 per KWh of electricity generated.[97] In case of Japan, the figures were 13 g /KWh for nuclear power according to the figures published in 2006. The UK Sustainable Development Commission report of 2006 gave a figure of 16 g/KWh for nuclear as compared with 891 g/KWh for coal and 356 g/KWh for gas.[98] These figures vouch for a favourable energy balance for nuclear power by any criteria and a very modest carbon dioxide output from the whole fuel cycle.

Now a strong case is being made for renewable based on the argument that if nuclear energy was safe then the nuclear industry would not require government guaranteed accidental liability protection as a condition for electricity generation. The issue of liability is touted as the major difference between atomic power and the renewable sources. Considering, "liability, cost, emissions, environmental justice, and weapon proliferation, renewable such as wind and solar photovoltaic are clearer safe, cheap, more climates friendly, more ethical, and less vulnerable to terrorism than nuclear fission".[99] Protagonists of renewable argue that nuclear industry's claim, on the one hand, that it is safe; and at the same time their demands for liability protection appears contradictory. The cost of fission electricity would also triple if market based, accident-liability insurance was to be taken into account. A Princeton University study published in *Science* has claimed that six of the nine renewable sources of energy already deployed could solve the global problem of climate change by 2050. The US government has said that wind energy from three states, Kansas, North Dakota and Texas could meet the US electricity needs. It is also argued that by 2015-18 the prices of solar photovoltaic will have grid parity and also will be cost competitive with all the electricity sources. The solar thermal is already fully competitive.[100]

Figure 2.12: Greenhouse Gas Emissions from Electricity Production

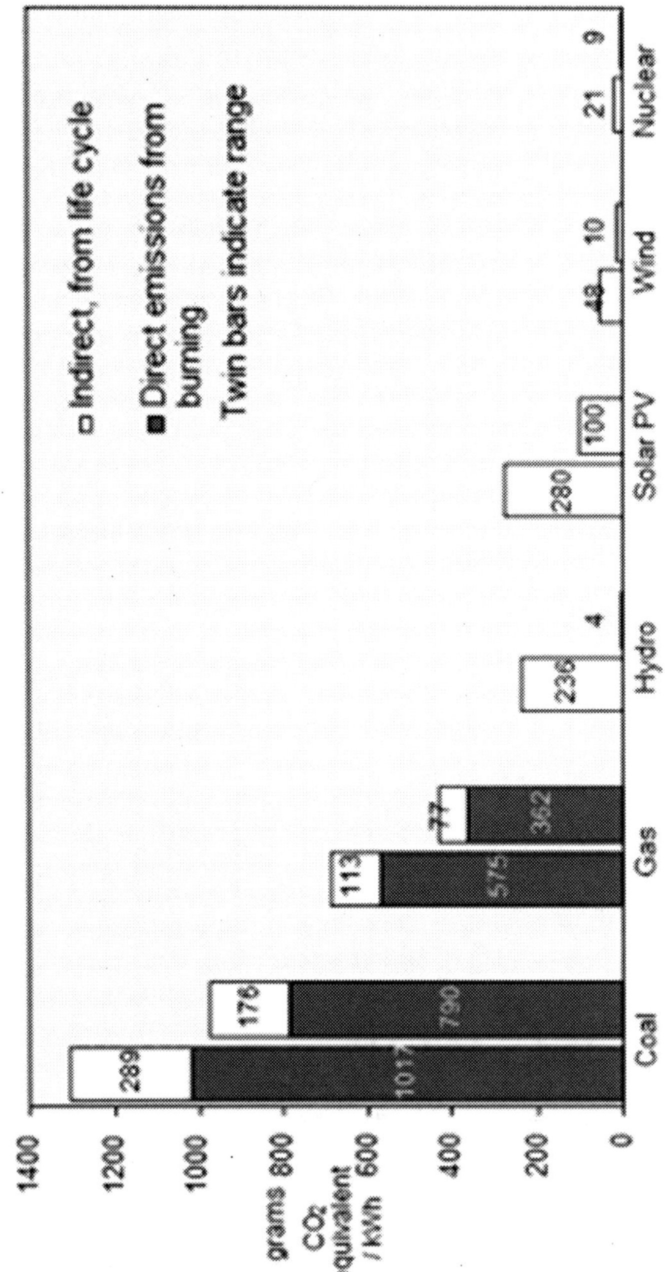

Source: IAEA, 2000.

Conclusion

Any additional nuclear based power generation capacity would displace thermal power generation (coal, oil and gas) and thereby reduce carbon emissions and help in the mitigation of the global warming and climate change and contribute, though in a limited way, towards ensuring the energy security, economic development and national security. Therefore, all steps should be taken to increase the nuclear power generation capacity in the country. All endeavours should be directed towards achieving targets. The nuclear energy can play a major role towards India's pursuit of energy security in the coming decades. However in view of the Fukushima nuclear crisis, concerted efforts would have to be made to improve the safety of nuclear reactors and further upgrade nuclear designs by introducing new safety features and gaining access to the fourth generation nuclear technologies.

India's energy deficit is so huge that all sources of energy should be tapped so that per capita electricity consumption can be raised from a dismal 778 KWh per annum to an average of minimum of 4000 KWh—that is considered essential for the human development.[101] There is very aggressive opposition to the nuclear power in India, from a section of the civil society groups and local population, who have opposed nuclear power vehemently, particularly after the Fukushima nuclear accident, for a variety of reasons. Steps should be taken to address genuine public concerns but the exaggerated fears and apprehensions spread by the motivated anti-nuclear groups must be ignored in pursuit of energy security and economic development. India should become more particular regarding the safety of nuclear reactors and make technological evaluations to ensure their safety. While expanding and building our nuclear power programme it is imperative that the entities entrusted with the job should take steps to build trust with public by being more transparent and should constantly be in communication with communities surrounding the power plant.

But given the miniscule share of nuclear power in the country's overall installed electricity generation, which at presents accounts for less than 3 per cent (2.25%), has led to serious scepticism about whether the nuclear energy departments will be able to achieve the very high and ambitious targets and future projections in view of the slow pace of the progress in building the reactors and the obstacles hindering future growth. This has raised a very pertinent question: Should not the government set more realistic targets as the previous records of missed goals must serve as a lesson for the future.

An appraisal of capabilities and constrains is essential. However, while implementing the nuclear programme a judicial mix of indigenously developed and imported nuclear reactors would help in achieving the targets.

Renewable energy sources such as wind, solar, hydro, biomass are the other options which India is committed to promote to meet the challenges of climate change and cut green house gases emissions from the coal based thermal generation of electricity. But the electricity generated from the renewable energy sources and the nuclear power cumulatively will constitute only a fraction of the share of the total electricity generation which would remain largely coal centric. Moreover, the renewables and the nuclear power have their own share of problems and vulnerabilities. Renewables are constrained to generate electricity at a large scale. They are, equally, highly unreliable as they are dependent upon so many other factors for continuous generation.

Grid connectivity poses another problem along with technological glitches. Solar and wind do offer some promise. Wind energy has already become competitive. Solar power generation is now the focus of the government as it is becoming competitive with other modes of power production and is likely to achieve cost parity with other sources of power. Concentrated Thermal Power (CSP) using solar power is fast emerging as a viable power generation method. Collaborative research with countries that have excelled in this area can be of immense value. The technology transfer option should also be explored wherever appropriate.

The promotion of nuclear energy should be viewed more as contributing to India's efforts to meet its energy deficit and enhancing of its energy security. Nuclear energy has the potential to supplement other sources of energy and foster economic development. It can help achieve India's goal of energy security and provide a low carbon energy option along with other renewable sources, and thus partly strengthen country's resolve to meet the challenges of climate change as spelled out in the National Action Plan for Climate Change. However, the over enthusiasm of developed countries seeking GHGs emissions reduction commitments from countries like India which may hinder its economic growth and development should be overlooked. But at the same time a determined resolve is needed to preserve the climate. Since environmental protection and economic development are compatible and not contradictory, both can be pursued with equal vigour and commitment.

NOTES & REFERENCES

1. Planning Commission, "Power and Energy", *Government of India*, URL: http://planningcommission.gov.in/sectors/index.php?sectors=energy, accessed on July 27, 2013.
2. Globally in 2010, there was an installed capacity of 17,000 MW, with Germany showing a growth of 40% which is expected to continue to grow in the coming years. This growth rate has been accompanied by the rapidly declining cost of production of solar power. The cost reduction has been possible because of economies of scale, technological advancements and emerging low cost manufacturing.
3. Anant Sudarshan and Ligia Noronha (2009), "Contextualizing India's Energy Security" in Ligia Noronha and Anant Sudarshan (eds), *India's Energy Security*, Routledge: London, pp. 5-6.
4. Planning Commission (2006), *Integrated Energy Policy: Report of the Expert Committee*, Government of India, August 2006, p. 54.
5. Ibid, *Integrated Energy Policy*, p.54
6. Ibid, *Integrated Energy Policy*, p.54
7. Ibid, Ligia and Noronha, p. 6-7.
8. Ibid. Ligia and Noronha, p. 6.
9. Ibid, *Integrated Energy Policy*, p.54.
10. Government of India, Planning Commission (2011), *Faster, Sustainable and More Inclusive Growth: An Approach to the Twelfth Five Year Plan (2012-17)*, October 2011, p.28, URL: http://planningcommission.gov.in/, accessed on July 26, 2013.
11. Arvind Gupta (2012), "MEA's role in energy security", *The New Indian Express*, August 2, 2012, URL: http://newindianexpress.com/opinion/article581090.ece, accessed on March 26, 2013. Oil public sector undertakings have invested over Rs. 64,000 Crores abroad. ONGC has 40 projects in 15 countries, IOCL has nine projects in six countries; OIL has 12 projects in eight countries, BPCL has 12 projects in seven countries, GAIL has four projects in two countries and HPCL has two projects in two countries, Petronet LNG Limited (PLL) imported 8.64 million metric tonnes of LNG at its Dahej terminal. It is also securing large contracts in Western Australia. The import of shale gas from the US is also being explored.
12. Ibid, *Integrated Energy Policy*, p.55.
13. Ibid, *Integrated Energy policy*, p.57.
14. The Energy Resource Institute (TERI) (2009), *National Energy Map for India: Technology Vision 2030*, TERI press, 2009, URL: http://bookstore.teriin.org/docs/books/psa/20, accessed January 27, 2013. The study found that the total commercial energy requirements across various scenarios would be as follows. In the Business As Usual (BAU) scenario the total energy consumption was estimated to increase by 7.5 times over the next 30 years from a level of 285 million tonnes of oil equivalent (Mtoe) in 2001 to 2123 Mtoe in 2031. Across the alternative growth scenarios the figures would be as follows: low growth (LG) economy growing at a low pace of 6.7 %, the commercial energy requirement would increase to 1579 Mtoe by 2031(5.9 % GDP Growth). In case of high growth (HG) of 8.6% it could go as high as 3351 Mtoe by 2031. However, the government's decision to increase the supply of hydro power, nuclear energy, and renewable resources would have a minor impact when compared with the total requirements of commercial energy by 2031.
15. Ibid, *National Energy Map for India: Technology Vision 2030.*
16. Ibid, *National Energy Map for India: Technology Vision 2030.* Measures to enhancing the end use efficiency; adopting advanced coal and gas based power generating technologies; increased exploitation of renewable and nuclear energy resources; and finally improving the efficiency in the transport sector. By adopting these strategies the commercial energy requirement can be reduced considerably over the next decades. It could be reduced in 2031, in a BAU

(Business as Usual) scenario from a level of 2123 Mtoe in the BAU to 1503 Mtoe. This can be achieved by adopting an integrated energy approach with a focus on the demand and supply alternatives.

17. Energy intensity is a measure of energy efficiency of nation's economy. It is calculated as units of energy per unit of GDP. High energy intensity indicates a high price or cost of converting into GDP. Low intensity indicates a lower price or cost of converting energy in to GDP.
18. Ibid, TERI, Technology Vision 2030, p.20.
19. Michael Kugleman (2011), "Integrating Energy Concerns into India's National Security Strategy", *IAGS Journal of Energy Security*, December 2011, URL: http://www.ensec.org/index.php?option=com_content&id=334:integrating-energy-concerns-into-indias-national-security-strategy&catid=121:contentenergysecurity1111&Itemid=386, accessed on July 30, 2013.
20. Ibid, Arvind Gupta.
21. Ibid, Anant Sudarshan and Ligia Noronha (2009), p. 8. The annual growth rate of the economy which has hovered around on 8% on an average for the last decade and the forecasts that the economy may grow even at a higher rate of growth of 8 to 10% in the coming years; the changing nature of the Indian economy, as its focus shifts from the agriculture to the service sector necessitating demand for infrastructure, and housing etc.; a growing population which is increasing at the rate of about 1.9% p.a. over the last two decades; and the increase in urban population from 25.5% to about 40% by 2030; the increasing demand for petrol and diesel due to the exceptional growth of the transport sector which constitutes about 50% of oil demand; with changing fuel choices. The demand for electricity is likely to rise significantly as currently biomass still is the major cooking fuel in rural India and less than 50 % of the rural households are electrified.
22. Ibid, Anant Sudarshan and LigiaNoronha, p.8.
23. Ibid, *Integrated Energy Policy*, p. xxiii.
24. Ibid, Ligia and Sudarshan, p.8.
25. Ibid, Michael Kugleman.
26. Ibid, *Technology Vision 2030*, p.22.
27. Ibid, *Technology Vision 2030*, p.22.
28. Ibid, *Technology Vision 2030*, p.23.
29. Ibid, *Technology Vision 2030*, p.23.
30. Ibid, Michael Kugleman.
31. International Energy Agency (2012), *Key World Energy Statistics 2012*, pp. 48-57, URL: http://www.iea.org/publications/freepublications/publication/kwes.pdf, accessed on August 10, 2013.
32. Ministry of Power (2013), *Power Sector at a Glance: All India* - As on June 30, 2013", *Government of India*, URL: http://www.powermin.nic.in/indian_electricity_scenario/introduction.htm, accessed on August 7, 2013.
33. Report of the Working Group on Power for Twelfth Plan (2012-17), *Executive Summary*, p.3, URL: http://planningcommission.nic.in/aboutus/committee/wrkgrp12/wg_power1904.pdf, accessed on August 10, 2013
34. Ministry of Statistics and Programme Implementation (2012), *Energy Statistics 2012*, Central Statistic Office, National Statistical Organization, Ministry of Statistics and Programme Implementation, Government of India, p.1., URL: http://mospi.nic.in/mospi_new/upload/Energy_Statistics_2012_28mar.pdf, accessed on August 15, 2013.
35. Some of the factors hampering the growth are: lack of competition; absence of benchmarks for different operational parameters; and absence of an independent Regulatory body; problem of land acquisition and rehabilitation; stringent legislation covering forest conservation and

environment management; non-transparency in the pricing; a complex distribution regime which is constrained by rail infrastructure which does not offer any flexibility; despite low import duty of 5% the imports of the thermal coal has been sluggish because of constraints of port capacity and the cost associated with multiple handling and inland transportation of imported coal; and the absence of appropriate legislation allowing entry of the private sector. See, *Integrated Energy Policy: Report of the Expert Committee,* August 2006, p.12.

36. "Centre to amend Coal Mines (Nationalisation) Act", *Infrawindow News Bureau,* May 15, 2012, URL: www.infrawindow.com/news/centre_to_amand_coal_mines_nationalisation_act_3276, accessed on June 13, 2013.
37. Ibid, *Integrated Energy Policy: Report of the Expert Committee,* August 2006, p.12.
38. Ibid, Report of the Working Group on Power for Twelfth Plan (2012-17), *Executive Summary,* p.6.
39. Coal bed Methane is a generic term for methane rich gas originating in coal –seams. It can be classified in three main categories: Coal-Mine Methane (CMM); Abandoned Mine Methane (AMM); and Virgin Coal Bed Methane (VCBM). A number of benefits have been attributed to CBM. For details see "CBM Exploration and Production", *Petroleum Federation of India,* March 2010, URL: http://petrofed.winwinhosting.net/upload/IAI/17-20mar10/CBMexploration.pdf, accessed on June 13, 2012.
40. Under Ground Coal Gasification (UGCG)/in-situ coal gasification is the most suitable technological method to make use of the low quality coal which has a very high level of ash.
41. Ibid, *Integrated Energy Policy: Report of the Expert Committee,* August 2006, p.12.
42. Ibid, *Energy Statistics 2012,* Central Statistics Office, p.1
43. The US Government's Energy Information Administration estimated in 2011 that India had just 63 trillion cu ft of technically recoverable shale gas resources as compared to China's 1,275trn cu ft.
44. Martin Adams (2012), "Empowering growth Perspectives on India's energy future: A report from the Economist Intelligence Unit, *The Economist,* p.8., URL: http://www.managementthinking.eiu.com/sites/default/files/downloads/Empowering_Growth.pdf, accessed on October 17, 2013.
45. Ibid, *Energy Statistics,* p.9.
46. Ministry of New & Renewable Energy (MNRE), *Achievements,* Government of India, URL: www.mnre.gov.in/mission-and-vision-2/achievements, accessed on February 4, 2013.
47. REC are generation based "certificates" awarded electronically, in demat form, to those who generate electricity from renewable sources such as wind, solar, hydro and biomass, in case they decide not to sell electricity at a preferential high tariff. These certificates are tradable on the exchange and are bought by "obligated entities" who are either specified consumers or electricity distribution companies. These obligated entities may either be required to purchase a certain quantum of green power or RECs. See, "Indian Renewable Energy Certificate market nosedives on poor demand", *Panchabuta,* August 29, 2012. REC is a market based mechanism to promote renewable sources of energy and development of electricity sector to meet the country's energy requirement and add to the energy security of the country. As the renewable resources are unevenly distributed in the country it would help encourage larger generating capacities in states which have renewable resources through this process of certification which is a market based instrument that can be traded. The REC seeks to socialise the cost of green energy development by facilitating RPO compliance in states with insufficient renewable resource capacity. And, RPO is the obligation required by the State Electricity Regulatory Commission (SERC) under the Energy Act 2003 to purchase a minimum quantity of renewable energy out of the total consumption in the area of a distribution licensee Whether these utilities fulfil their obligations would be supervised by compliance auditors who would check whether the participants in the scheme have

complied with the REC requirements. Renewable Energy Certificate Registry of India, *About REC*, URL: http://www.recregistryIndia.nic.in/index.php.general/public/AboutREC, accessed on December 11, 2013.

48. Ibid Renewable Energy Certificate Registry of India.
49. Indian Engineering Exchange, *About Renewable Energy Certificates (REC)*, URL: http://www.iexindia.com/rec.htm, accessed on December 12, 2013.
50. For more details on the functioning of REC and RPO see KMPG-India (2011), "*The Rising Sun: Point of View on the Solar Energy Sector in India*", May 2011, p.27, URL: http://www.kpmg.com/IN/en/IssuesAndInsights/ThoughtLeadership/The_Rising_Sun_full.pdf, K. Soone, Minaxi Garg, and Satya Prakash (2010), "*Renewable Energy Certificate Mechanism in India*", URL: http://posoco.in/papers/Renewable%20Energy%20Certification%20Mechanism%20in%20India_2010.pdf, accessed on December 12, 2013
51. Ibid S.K. Soone et.al.
52. For recent developments in REC trading see, "Nov 12 REC Trading: Traded Volume falls10% of supply at floor price", *Climate Connect*, November 30, 2012, URL: http://www.climate-connect.co.uk/Home/?q=November%202012%20REC%20Trading%3A%20Traded%20volume %20falls%20to%2010%25%20of%20supply%2C%20cleared%20at%20floor%20price, accessed on October 15, 2013; "November REC Issuance dropped by a drastic 40% since October", *Climate Connect*, November 29,2012, URL: http://www.climate-connect.co.uk/Home/?q=November%20REC%20issuance%20dropped%20by%20a%20drastic%2040%25%20since%20October%20, accessed on October 15, 2013
53. "Make renewable energy certificate trading simple to encourage corporate", *Times of India*, July 14, 2012.
54. REConnect Energy Solutions (2012), "CERC Declares New REC Floor and Forbearance Price for FY 2012-2017", August 25, 2011, URL: http://reconnectenergy.com/blog/2011/08/cerc-declares-new-rec-floor-and-forbearance-price-for-fy-2012-2017/, accessed on December 13, 2012. The prices at which REC are being traded seem to disappoint the industry as the certificates are changing hands at the minimum prices fixed by the government. The average price at the Indian Energy Exchange LTD. which accounts for almost 90.6% of the trade was around Rs1500 for each REC which is the floor price. On the power Exchange of India LTD, the average price was Rs 1,555. Moreover, the increase in volume was also not very heartening as the major buyers were the same such as Tata Power, Reliance Infra, and Torrent. Since these entities have already met most of their RPO obligations they are not likely to turn-up next month. However, the biggest disappointment among the renewable energy producers is that no state owned "obligated entities" such as Public Discom (electricity Distribution Company) and large captive plants, have come forward to buy the REC certificates. This is because of the lack of enforcement of their obligations by the government. See, "Indian Renewable Energy Certificate market nosedives on poor demand", *Panchabuta,* August 29, 2012. The slump in the REC trading and non-compliance of RPO by many state entities as they are not buying these certificates has led to a situation where the states with abundant renewable potential such as Rajasthan, Karnataka, and Tamil Nadu are not keen to develop more renewable energy than required by their respective RPO.
55. Ibid, *MNRE - Achievements.*
56. Some of these entities are: Orb energy; simpanetworks; Husk Power System; Barefoot Power; SELCO; Auroville Renewable Energy; Nuru-Energy to Empower and Schneider. Simpanetworks, "*Our Solution*", URL: simpanetworks.com/our-solution, accessed on December 17, 2013; Husk Power System, URL: www.huskpowersystem.com, accessed on December 17, 2013; Sushma UN (2012), "Orb Energy devices solar lighting kit for rural areas", *The Hindu Business Line,* March 3, 2012, URL: http://www.thehindubusinessline.com/industry-and-economy/article2957907.ece, accessed on December 17, 2012.

57. Varun Mittal (2012), "Solar Solutions: Greening telecom in India", *Solar Novus Today*, July 18, 2012, URL: www.solarnovus.com/index.php?view=article&catid=77%3Acase-studies-and-solar-solutions&id=, accessed on December 18, 2013.
58. Ibid, *MNRE Achievements.*
59. "Outlook on Indian Renewable Energy – 2012", *Panchabuta,* p.4., URL: http://panchabuta.files.wordpress.com/2012/11/panchabuta_india_renewable_energy_outlook.pdf, accessed on December 28, 2012.
60. Ibid, *Outlook on Indian Renewable Energy – 2012*, p.4.
61. Feed in Tariff: Respective State Electricity Regulatory Commissions have adopted a cost plus approach wherein the cost associated with the setting up of a wind power plant; such as capital expenses and operational expenses, are considered on the basis of discussions with the stakeholders. Tariff is calculated as per the levelised cost that is achieved through these assumptions. A small margin is added to enable the developer to make a profit. Another factor that is taken into account is the capacity utilisation factor (CUF). Different states have different wind regimes. This affects the electricity generation and consequently the revenue. In order to take into account this difference of wind density a mechanism called "zone based "tariff has been designed. In this system regions with lower wind density are offered a higher tariff while zones with higher wind density are offered lower tariff. This is done to ensure that tariffs for all the zones remain the same and the variance due to wind energy regime is minimised.
62. Under the Accelerated Depreciation (AD) scheme wind farm developers have been given tax benefits in the form of 80% AD on their assets motivating wind farm developers to enter the wind energy generation. The government has provided dual incentives of energy generation for self consumption and tax benefits under the AD Scheme. The AD benefits have been now withdrawn with effect from the financial year 2012-13 i.e. from April 1, 2012.
63. Generation Based Incentives launched in 2009. The developer had the choice to either opt for AD or GBI system. The GBI mechanism offered a GBI of Rs.0.50 per KWh with a predefined cap of Rs. 62.5 lakh per MW for the capacity installed. The incentive was available till the end of last financial year i.e. all projects commissioned before March 31, 2012. GBI were available till 4000 MW of capacity were allocated under the scheme till March 31, 2012. However, till this date less than 2000 MW capacity has been allocated and the deadline has since passed. As per media reports MNRE has recommended a revised GBI of Rs.0.82 per KWh with the total capacity available to be allocated pegged at 13,500MW. As there are very remote chances of AD scheme being reintroduced the Wind farm developers will have to depend on GBI. See, *Outlook on Indian Renewable Energy – 2012*, p.8.
64. Ibid, *Outlook on Indian Renewable Energy 2012*
65. Sujay Mehdudia (2010), "Manmohan Singh Launches 'Solar India', *The Hindu,* January 12, 2010.
66. Council on Energy, Environment and Water, and Natural Resources Defense Council (CEEW and NRDC), "*Concentrated Solar power: Heating Up India's Solar Thermal Market under the National Solar Mission*", September 2012, p.6.
67. See, Council on Energy, Environment and Water (CEEW), and National Resources Defense Council(NRDC*), Laying the Foundation for a Bright Future: Assessing Progress under Phase 1 of India's National Solar Mission, Interim Report*, April 2012, pp.11-13.
68. Ibid, *CCEW & NRDC.*
69. Ibid, *CCEW & NRDC.*
70. European Commission Research and Innovation, "*Concentrated Solar power*", URL: http://ec.europa.eu/research/energy/eu/research/csp/index_en.htm, accessed on December 21, 2012.
71. Non-Recourse Finance means a loan where the lending bank is only entitled to repayment

from the profits of the project the loan is funding, not from other assets of the borrower. Recourse loans are loans that allow the lender to come after you in case you default. You can contrast recourse loans with non-recourse loans, which create more risk for lenders. These types of projects are characterized by high capital expenditures, long loan periods, and uncertain revenue streams. Analyzing them requires a sound knowledge of the underlying technical domain as well as financial modelling.

72. Direct Normal Irradiance (DNI) is the amount of solar radiation received per unit area by a surface that is always held perpendicular (or normal) to the rays that come in a straight line from the direction of the sun at its current position in the sky. Typically, you can maximize the amount of irradiance annually received by a surface by keeping it normal to incoming radiation. This quantity is of particular interest to concentrating solar thermal installations and installations that track the position of the sun. 3-TIER Support, URL: http://www.3tier.com/en/support/solar-prospecting-tools/what-direct-normal-irradiance-solar-prospecting/, accessed on December 24, 2012.
73. Aerosols are extremely-fine liquid droplets or solid particles those remain suspended in air as fog or smoke. These non-gaseous microscopic bits (with radius from one-hundredth to one ten-millionth of a centimetre, or 100 to 0.001 micrometer) are produced from natural and artificial sources, and usually includes mineral dust, pollen, and sulphuric acid, URL: http://www.businessdictionary.com/definition/aerosol.html#ixzz2FxKnuGt2, accessed December 24, 2013.
74. All these points have been taken from Council on Energy, Environment and Water, and Natural Resources Defence Council (CEEW and NRDC), *Concentrated Solar power: Heating Up India's Solar Thermal Market under the National Solar Mission*, September 2012, pp.15-16.
75. Power Ministry, "Power Sector at a Glance - All India", *Government of India,* URL: http://powermin.nic.in/indian_electricity_scenario/introduction.htm, accessed on June 26, 2013.
76. Tarapur Atomic Power Station (TAPS), Maharashtra, 4 Reactors, 2 BWR and 2 PHWR; Rajasthan Atomic Power Stations (RAPs) 6 Reactors, all PHWRs; Madras Atomic Power Station (MAPS), 2 Reactors, both PHWR; Kaiga Generating Station (KGS), Karnataka, 4 Reactors all PHWR; Narora Atomic Power Station (NAPS), Uttar pradesh,2 Reactors, both PHWR; Kakrapar Atomic Power Station (KAPS), Gujarat, 2 Reactors both PHWR. Total Units, 20, Total Nuclear Power Plant Capacity, 4780 MW. See, Nuclear Power Corporation of India (NPCIL), URL: www.npcil.nic.in/main/All ProjectOperationDisplay.aspx, accessed on February 12, 2013. Kudankulam-1 VVER-1000 linked to the national grid on October 21, 2013.
77. Kudankulam Atomic Power Project, Tamil Naidu, 2 Reactors 1000x2MW, PWR; Kakrapar Atomic power Project, 2 Reactors, 700x2MW; Rajasthan Atomic Power Project (Rawatbhata) 2 Reactor, 700 MW×2MW; Fast Breeder reactor, Kalpakam Tamil Naidu, 500 MW. See NPICL, URL: www.npcil.nic.in/main/ConstructionDetails.aspx.ReactorID =77, accessed on October 12, 2013.
78. Kudankulam 4 Reactors PWR 1000x4; Jaitapur 4 Reactors EPR 177x4; Kaiga 2 Reactors PHWR 77x2; Gorakhpur 2 Reactors PHWR; Chukka 2 Reactor PHWR 700x2; Bheempur, Madhya Pradesh, 2x700 PHWR; Bans Wada, Rajasthan, 2x700 PHWR; Kalpakam, 2x500 FBR. Total 18 Reactors are planned with 15,700 MW. See, World Nuclear Association, URL: www.world-nuclear-org/info/inf53.html, accessed on February 12, 2013.
79. Kudankulam 2x1000 PWR –AES; Gorakhpur 2x700 PHWR; Rajauli (Bihar) 2x700PHWR, 2x1000 PWR; Jaitapur 2x1700 EPR, 2x 500 FBR?, AHWR 300 ;Jaitapur PWR-EPR1600; Markandi (Pati Sonapur) Orissa, PWR 6000MW; Mithi Virdi Gujarat Saurashtra, 4xAP1000, 1250 MW; K. See, World Nuclear Association, URL: www.world-nuclear-org/info/inf53.html, accessed February 12, 2013; Kovvada (Srikakulam Andhra Pradesh),

4xESBWR, 1350 & 1550 MW; Nizampatnam (Guntur, AP), 1-6, 6x1400; Haripur 8x VVER 1200, PWR; Pulivendula (Kadapa AP) 2x1000 BWR& 2x700 PHWR;Chutka2x1400 PHWR; Mithi Virdi 2xAP1000, 1250 MW; Kovvada ESBWRx2, 1400x2. Total 39; See, World Nuclear Association, URL: www.world-nuclear-org/info/inf53.html, accessed February 12, 2013.

80. Information was given to the India Lok Sabha by Shri V. Naranaswamy, Minister of State in the Ministry of Personnel, PG &Pensions and in the Prime Minister's Office, Government of India, on August 9, 2012, URL: http://pub.nic.in/newsite/erelease.aspx?relid=85902, accessed on October 18, 2013.
81. R.B. Grover and N. Nagaich (2012), "Nuclear Plans: Augmenting Generation and Indigenous Manufacturing Capacity", *Power Line,* January 2012, URL: http://npcil.nic.in/pdf/news_09feb2012_01.pdf, p.94, accessed on October 12, 2013.
82. Ibid, R.B. Grover and N. Nagaich, p. 94.
83. World Nuclear Association, "*Comparative Carbon Dioxide Emissions from power Generation*", URL: www.world-nuclear.org.education/comparativeco2.html, accessed on February 13, 2013.
84. Dianne L. Durante (1991), "Nuclear Power: How it works and why it's Better than other fuels", *Forgotten Delights,* URL: www.forgottendelights.com/essay/Nuclear power.htm, accessed on February 20, 2013.
85. Ibid, Dianne L. Durante.
86. Charles Forsberg (2011), "Mutually assured energy independence", *Bulletin of Atomic Scientists,* April 28, 2011, URL: www.thebulletin.org/print/web-edition/rountables/nucear-enrgy-different-other-energy-sources?order=asc, accessed on October 6, 2013.
87. Ibid, Charles Forsberg.
88. Tony Pietrangelo (2011), "Low carbon, low cost electricity 24/7" *Bulletin of Atomic Scientists,* July 28, 2011, URL: http://thebulletin.org/nuclear-energy-different-other-energy-sources/low-carbon-low-cost-electricity-247, accessed on October 7, 2013.
89. WNA Report, Comparison of Life Style Green house Gas Emissions of Various Electricity Generation Sources, URL:http://www.worldnuclear.org/uploadedFiles/org/WNA/Publications/Working_Group_Reports/comparison_of_lifecycle.pdf, accessed on April 22, 2013.
90. Ibid, Tony Pietrangelo.
91. Ibid, Tony Pietrangelo.
92. Ibid, Tony Pietrangelo.
93. Kristin Shrader-Frechette,: Cheaper, safer alternatives than nuclear fission', in Charles Forsberg, *Bulletin Atomic Scientists,* It is estimated that during the last 50 years the US has provided a subsidy of $165 billion to the commercial nuclear sector as compared to $5 billion to the renewable sector—solar and wind combined. And if we count both direct and indirect subsidies the total subsidies for the commercial fission amount to almost $1 trillion at the rate of $20 billion annually which is almost 200 times those for wind and solar. According to credit rating agencies the market based costs of nuclear energy was almost 15 cents per kilowatt-hour and was rapidly rising while the market based cost for the wind was about 3.4 cents, which is five times less than those of the nuclear electricity. Sometimes dropping to 1 to 2 cents per kilowatt hour. The high costs as compared to the renewable are likely to adversely impact the share of nuclear energy in the total global energy production. It may come down to about 9 per cent from its present share of 14 per cent. The globally installed capacity of wind power is likely to reach 380 gig watts by 2013 and there are prospects of its rising in future while the nuclear installed capacity was around 375 gig watts and is declining.
94. Ibid, Tony Pietragelo.

95. Knowledge Wharton, "US Energy Policy after Japan: If not Nuclear, than what?", URL: knowledge.wharton.upenn.edu/printer_friendly.cfm?articleid_2743, accessed on February 20, 2013.
96. Ibid, Tony Pietrangelo,
97. World Nuclear Association, "*Comparative Carbon Dioxide Emissions for power Generation*", URL: www.world-nuclear.org.education/comparativeco2.html, accessed on February 13, 2013.
98. Ibid, Comparative Carbon Dioxide Emissions for power Generation.
99. Kristin Shrader-Frechette (2011), "Cheaper, safer alternatives than nuclear fission", *Bulletin of Atomic Scientists,* August 19, 2011, URL: http://thebulletin.org/nuclear-energy-different-other-energy-sources/cheaper-safer-alternatives-nuclear-fission, accessed on October 10, 2013
100. Ibid, Kristin Shrader.
101. Andrew Teller, "Correlation is not causation", *ENS News,* Issue 25 (Summer), July 2009. URL: https://www.euronuclear.org/e-news/e-news-25/listening.htm, accessed on February 22, 2013. At the global level, the forecasts of the level the future energy requirements can be made only if human development needs are taken in to account and this can be best based on the use of human development index which shows a measure of human wellbeing. Alan Pasternak has assessed that minimum of 4000KWh is the threshold for this. The reason for putting this figures according to him is that no country with an electric consumption below 4000KWh has an HDI above 0.9 and barring four cases (South Africa, Saudi Arabia, Russia and South Korea) all countries consuming more than this value per capita have an HDI greater than 0.9.

CHAPTER 3

Public Concerns over Nuclear Power in India

Kapil Patil

Introduction

The Fukushima crisis has severely affected the global public acceptance of nuclear power. The widespread media coverage of the Fukushima Dai-ichi nuclear reactor meltdown has generated fears relating to catastrophic accidents and rekindled the debate on the safety and risk parameters of nuclear energy. The echoes of the Fukushima crisis strongly resonated in India which witnessed agitations by people and civil society groups against the new nuclear power plants (NPPs) planned in different part of the country. Post-Fukushima, the public support for nuclear energy in India appears to be declining. The societal consensus, that the benefits of nuclear energy outweigh the risks, also seems to be breaking down following the growing public opposition to nuclear energy projects. The protesting public at various nuclear plant sites such as Kudankulam, Jaitapur, Gorakhpur, Kovvada, and Mithi Vardi, have expressed concerns ranging from lack of information about the nuclear power to tangible concerns such as losing land/water rights, means of livelihood, nuclear safety and environmental destruction, etc.[1] The civil society groups on the other hand resorted to legal activism and filed petitions in the Supreme Court seeking ban on the construction and commissioning of new NPPs and uranium mines in the country.[2] A section of public intellectuals and the scientific community strongly contested the nuclear safety review conducted in the aftermath of Fukushima nuclear crisis and demanded a separate inquiry into the safety aspects of India's nuclear installations by independent group of experts. The anti-nuclear campaigns therefore exerted immense pressure on policy-makers to review the country's nuclear energy policy.

The public perceptions of the severity of risks associated with nuclear power, however, are quite different from the actual risk posed by the NPPs. Public opinion on nuclear energy differs considerably from the scientific evidence on safety and risk parameters of NPP's. Numerous studies have shown that how people perceive risks bear no particular relation to the actual hazards ranking of nuclear power.[3] While the scientific facts on nuclear energy are often unknown to common people; the studies on public attitudes have identified various factors that shape people's perceptions of risks associated with nuclear power. The Fukushima crisis in Japan—which posed a serious challenge to the future of nuclear energy—has highlighted the profound impact public opinion can have on policy-making. While harnessing nuclear power to meet the growing energy needs, it is essential that public concerns over risks emanating from the NPPs be addressed and minimised.

In the light of changing public perceptions of nuclear energy post Fukushima; this chapter seeks to situate public concerns over nuclear energy in the broader conceptual framework of risk assessment and public attitudes towards nuclear installations. Second, it seeks to identify the specific concerns expressed by common people during the public hearings at new NPP and Uranium mining sites. It takes into account the concerns expressed by independent experts working in the areas of environmental conservation, sustainable development and public health. Finally, the chapter aims to critically evaluate the risk management measures that the Atomic Energy Regulatory Board (AERB), Nuclear Power Corporation of India Limited (NPCIL), and the Department of Atomic Energy (DAE) have undertaken to address safety related concerns and inform the lay people about the various risks arising from nuclear power generation. Nuclear energy in India is widely perceived as an important constituent of the overall energy mix and a source of long term energy security. However, the viability of nuclear energy as a source of long term energy security depends critically on minimising the various risks arising from nuclear installations and improving its wider public acceptance in the country.

Nuclear Energy and Public Attitudes

The public perceptions of risks associated with nuclear installations have been an important area of social science research for the past three decades. The environmental movements of the 1960s and 1970s in Europe and USA were centred around the atmospheric testing of nuclear weapons and radioactive waste dumping in the oceans. The Three Mile Island accident in 1979 and more importantly the Chernobyl in Ukraine, 1986, had strongly affected the public acceptance of nuclear power. The Chernobyl accident manifested for the first time, the serious trans-boundary health and environmental consequences of radioactive contamination. Post-Chernobyl, many nuclear power producing countries in Europe witnessed strong opposition to accept array of risks arising from nuclear facilities. The studies on people's attitude towards nuclear energy after Three Mile Island and Chernobyl accidents have offered useful insights into people's perceptions of risks associated with NPPs and radioactive waste storage facilities.

Paul Slovic has been the pioneering risk analyst who attempted to understand public perceptions of nuclear power.[4] A study conducted by Slovic found that people judge the benefits from nuclear power to be quite small and the risks to be unacceptably great.[5] In his view, people's fears about nuclear risks are unknown, uncontrollable and inequitable, whereas the industry experts rank nuclear power in the bottom tier of risks when compared to other hazardous technologies.[6] Slovic noted that nuclear power is a special type of risk in the public imagination, sometimes creating extreme perceptions relative to other types of risks. Slovic's key argument is that lay risk perceptions are often much richer than that of the industry experts and sometimes reflects concerns that are typically omitted from experts risk assessments. In other words, unlike experts with narrower technical determinations of risk; the risk perceptions of lay people are often much widespread, and difficult to avoid. An important aspect of risk perception studied by researchers is the political affiliation of public groups in a given territory. Costa-Font et al in their study reveal that political orientations and beliefs play a determinant role in shaping the public attitudes towards nuclear power than the knowledge or perceived knowledge about the consequences of nuclear energy use.[7] Using survey data from a 2005 Eurobarometer of UK citizens; researchers claimed that views on nuclear power are driven by an instinctive political position that is often deeply held by citizens. Individuals with left leaning views are more likely to hold that nuclear power is a highly risky technology as compared to those with moderate or right leaning political views.[8] Kasperson et al. in their study have noted that environmentalists as a group are substantially more critical than the public in their assessment of the safety of nuclear energy. Not only do they see accidents, nuclear explosions, and escape of radioactivity as major problems, but also express grave concern over plutonium production.[9]

Another important set of studies attempted to explore the link between risk communication efforts of nuclear authorities and risk perceptions of lay people. While the public awareness campaigns by government agencies and certain NGO's are aimed at increasing individual's knowledge about nuclear energy; the studies have concluded that scientific knowledge about nuclear energy or the availability of technical information, has no direct impact on lay risk perceptions. As Costa-Font suggests, in some instances, a general knowledge of the nuclear technology can actually decrease support for nuclear power.[10] The media coverage often evokes emotional reactions to nuclear accidents and plays an important role in influencing risk perception of relatively uninformed people. Nevertheless, the studies highlight that instinctive political beliefs determine to a greater extent associated risks perceptions and policy support towards nuclear energy. In addition to the importance of political orientations, trust in institutions (regulatory), and concerned authorities significantly alters the public acceptance of nuclear energy. Whitfield et.al in their study of institutions find a strong and consistent relationship between levels of risk perception and levels of trust in institutions, information, individuals or other objects of trust.[11] They argue that attitudes toward nuclear power are a function of perceived risk, and that both attitudes and risk perceptions are a function of values, beliefs, and trust in the institutions that influence nuclear policy.[12] Also Freudenberg shows, citizens desire a process (for siting nuclear installations) that is truly equitable, unbiased and fair – particularly in the sense of being fair to the values that are most important to the specific citizens involved, and detest any form of secrecy or exclusivity on part of the nuclear authorities.[13] Maintaining credibility in the public's eyes is therefore critical for nuclear authorities.

The extant literature on public perceptions offers many useful insights into the dynamics of people's assessment of the risks arising from nuclear installations. An important finding from the research as surmised by Slovic is that "attitudes toward nuclear power are conditioned by the interplay of psychological, social, cultural, historical, and political factors that will not easily be changed by public information or educational campaigns."[14] This is particularly critical for redressing public concerns vis-à-vis the new nuclear power projects. The studies also underscore the role of factors like availability of technical information, fears of social & environmental hazards and safety concerns other than factors like trust in nuclear authorities, and political orientation, in moulding the risks perceptions of lay people. The growing public opposition to nuclear energy in India is driven by many of the aforementioned factors, individually or in combination. The following sections aim to understand various concerns that contribute to risk perceptions of lay people in the Indian context.

Public Concerns over Nuclear Energy in India

Public concerns regarding nuclear power in India have assumed immense political significance in the light of growing public opposition to new nuclear energy projects. India's nuclear energy programme which began expanding earnestly in 1980s and 1990s has enjoyed substantive social and political support in the past. However, in recent years, nuclear projects have come under intense criticism by various environmental groups which began raising questions about the environmental impact of nuclear mines and power plants in the country. Environmental groups like Greenpeace have been spearheading the agitations against the development of various nuclear energy projects in the country. At the grass-root level, communities of farmers and fishermen view nuclear projects as a threat to their agricultural farmlands and fishing grounds. The socially marginalised tribal communities in the states of Meghalaya and Andhra Pradesh have also opposed the opening up of the new uranium mines in these regions.

The Fukushima crisis in Japan has further intensified public opposition to nuclear power in India. Nuclear safety concerns generated by Fukushima crisis, have transformed the isolated opposition over individual nuclear energy projects into a coordinated national campaign by certain civil society groups. The coordinated campaign not only highlighted various concerns related to nuclear power but also forced the government to review country's nuclear energy policy. The public opposition to new NPPs and uranium mining projects can be broadly classified into three major categories: radiation & safety concerns; environmental concerns – (impact on agriculture, horticulture, & marine eco-system) and socio-economic concerns. These categories are not meant to be mutually exclusive and overlaps are possible among factors falling under these broad categories that contribute to risk perceptions of people in the Indian context.

I) Nuclear Safety and Radiation Related Concerns

Nuclear energy is often viewed as a scientific discovery that is innately destructive in nature. The accidents in the nuclear industry are often equated with the atomic bombing of Hiroshima or Nagasaki which caused thousands of deaths. There have been three major reactor accidents in the history of civil nuclear power—Three Mile Island (1979), Chernobyl (1986) and Fukushima (2011). The Three Mile Island meltdown was successfully contained without any harm to the public; the Chernobyl disaster involved an intense fire and radioactive fallout due to lack of reactor containment; and the Fukushima core meltdown severely tested the integrity of the reactor containment, allowing release of radioactivity.[15] These accidents continue to spark debates on safety and risk parameters of nuclear energy. Since the Chernobyl accident, significant efforts have been devoted to make nuclear reactor technology safer. With several advances in reactor technology, and its promise to deliver clean energy at a competitive tariff, the nuclear industry had begun to slowly recover from the dreadful memories of Chernobyl accident. The events at Fukushima however, cast fresh doubts over the safety of NPPs. It led the critics to argue that nuclear accidents will be more frequent with varied origins, progressions, and impacts and the next nuclear accident will probably be caused by a different combination of initiating factors and failures.[16] The critics also rebuked the nuclear industry for failing to take into account various beyond-design based scenarios that could trigger a major systemic failure in the nuclear plant.

In addition to the possibility of catastrophic accidents, the fear of accidental release of radiation from NPPs remains a grave concern causing public opposition to nuclear energy. The health effects of excessive exposure to radiation can be potentially hazardous, but the regulated applications of ionising radiation have been immensely helpful in treating cancer disease. Notwithstanding its medical applications, the invisibility of radiation induces fear in the public mind. Radiation related concerns are often aggravated because of the deaths attributable to the Chernobyl accident and instances of cancer allegedly caused by the massive radioactive outfall. Although, the scientific debate on impacts of low levels of radiation on human life remains inconclusive, the fear of radiation from NPPs has been a major reason for public opposition to nuclear energy projects in India.[17]

There have been considerable apprehensions among certain civil society groups regarding the impact of NPPs as a result of studies conducted by independent experts, which suggested that there is a significant rise in diseases and disorders in the population around various NPPs and uranium mine sites.[18] A study conducted by a group known as 'Indian Doctors for Peace and Development' at the Jaduguda uranium mines concluded that more children with congenital deformities are being born to mothers and congenital defect as a cause of death of a child is also high among mothers living near uranium mining operations area; and cancer as a cause of death is more common in villages surrounding uranium operations.[19] While these studies have been widely cited to demonstrate the ill-effects of nuclear plants in India; the scientists in the Department of Atomic Energy have rejected these findings on lacking expert peer-review.[20] The DAE on the other hand, has contended that there has been no measurable increase in the incidence of cancers or any other morbidity patterns in the population as a result of the operation of NPPs in India.[21] The studies and surveys conducted by the DAE to assess the impact of radiation have shown that there is no significant increase in the radiation exposure for NPP workers, their families and population around the nuclear plants. Also the incidence of congenital deformities and cancer among workers and people living in the vicinity of nuclear installations are similar to that of the national average.[22] The fear of radiation and its impact on human society, however largely drives the opposition to nuclear energy projects.

During the public hearing at Jaitapur, local people opposed the DAE's decision to set up six French designed European Pressurised Reactors (EPRs) at the proposed site. The local people have expressed concerns over the untested EPR reactor design and the NPCIL's inexperience of building and operating such reactors in India.[23] The protestors at Kudankulam site too have questioned the safety of the Russian VVER reactors especially after the Fukushima crisis. Despite repeated assurances from various technical expert groups; the general public remains sceptical about the safety of the Russian rectors. The expert committee appointed by the central for allaying fears about the safety of the VVER reactors, in its report concluded that:

> The Kudankulam Nuclear Power Plant (KKNPP) is designed and engineered to the state-of-the-art nuclear reactors in line with the current

> international safety requirements and principles. The design of structures, systems and components, and the ample margin in the elevation of components provided a high level of safety against such events.[24]

The expert committee's assurances however, failed to break the deadlock and stop the protestors from launching new wave of protests. The protestors at Kudankulam site have criticised the Atomic Energy Regulatory Board for its lack of institutional independence and its inability to force the (NPCIL) to undertake stringent safety measures. The committee set up by the AERB following the Fukushima accident 'to review the safety of Indian NPPs against external events of natural origin', suggested certain additional safety measures for Kudankulam nuclear plant. The AERB's decision to allow the NPCIL to load fuel into Kudankulam reactors before the completion of additional safety measures drew widespread resentment from peoples' representatives. The former chairperson of AERB, A. Gopalakrishnan, termed the AERB's approval for fuel loading, "a total volte-face...and contrary to the spirit and recommendations of AERB post-Fukushima safety evaluation committee.[25]

The protestors alleged the lack of transparency on part of nuclear regulators and operators regarding the safety aspects of NPPs in India. The allegations related to lack of transparency correspond to general allegations of corruption and inefficiency directed against most government functionaries in India. A majority of the people believe that a Fukushima-like accident in India would be far more devastating due to poor enforcement of emergency planning zones and the all pervasive shortcomings of emergency preparedness measures. While the DAE may have reasonably safe track-record in management of nuclear installations; the allegations about lack of transparency in nuclear safety and non-autonomous structure of the regulatory agency have dented public faith in safety of nuclear energy projects in the country.

II) Nuclear Energy and Environmental Concerns

Nuclear energy is widely believed to be the best panacea for problems of climate hazards. The proponents of nuclear energy argue that, nuclear power can make an important contribution towards reducing greenhouse gases while producing energy in large quantities thereby, mitigating the impact of climate change.[26] The setting up of a nuclear power plant and fuel cycle facilities nevertheless, does impinge to a certain extent, on the surrounding environment and population. The construction of any mega energy project viz. nuclear power plants, thermal or hydro-electric power stations involves activities like digging, excavation, and cutting of trees which results in the loss of local biodiversity and animal habitats. The public opposition to mega power plants whether thermal or nuclear is therefore, mainly due to the immediate impact on the surrounding environment. The problems in the public acceptance of mega developmental project are further aggravated by the perceived violations of environmental laws by federal agencies in India.[27]

The reasons against the setting up of new nuclear energy installations are formally expressed through public hearing processes of the Environmental Impact Assessment (EIA) in India. The Environment Impact Assessment Notification of 1994 lists 'nuclear power projects' and 'nuclear fuel processing facilities' as requiring environmental clearances.[28] The Department of Atomic Energy (DAE) is mainly responsible for commissioning the EIA studies for new nuclear projects. The EIA process for nuclear installations has drawn serious criticism from environmental groups and local public. The EIA report for proposed Kudankulam nuclear plant III&IV, drafted by National Environmental Engineering and Research Institute (NEERI) has been criticised by Ramanna and Rao for stating smaller values of atmospheric radioactive discharges than the corresponding figures for VVER-1000 reactors operational in other countries.[29] They argue that for KKNPP 3 to 6, the Iodine-131 discharge for each reactor is given as 2,48,000 becquerels per day, which is six times less than the discharge rate for Khmelnitsky 1 reactor and 7.4 times less than the discharge rate for the Rovno 3 reactor, both of which are located in Ukraine.[30]

Various types of radioactive wastes are generated at a nuclear plant. The release of low level radioactive elements from the NPPs has been a cause of concern for environmental groups. At Kudankulam, the people's representatives contend that the KKNPP in its normal course of action would emit Iodine 131, 132, 133, Cesium 134, 136, 137 isotopes, strontium, tritium, tellurium and other such radioactive particles into the atmosphere with serious health consequences.[31] Radioactive releases from NPPs during its operational and de-commissioning phase require stringent monitoring and control measures to minimise their release into the atmosphere. The releases of radionuclides in liquid and gaseous phases from the NPPs, are therefore, made on the basis of the ALARA principle (as low as reasonably achievable). Similarly, the liquid level waste generated at the NPP is treated as per the prescribed regulatory limits through certified treatment systems.

Both the common people and environmentalists are substantially more concerned about the radioactive spent fuel generated from the NPPs. The entry of nuclear wastes into the bio-sphere is a major source of public opposition to nuclear energy. For many environmental groups, the production of plutonium – a radionuclide of extreme longevity—symbolises the unique risks posed to the environment and the personnel involved in re-processing facilities.[32] The lay people and the civil society groups in India, generally appear unwilling to accept the claims of DAE scientists on issues of nuclear waste disposal causing trust deficit between the nuclear industry and people.

Another important environmental concern of the people residing in areas surrounding the NPPs pertains to thermal discharges from the reactor operations. The unregulated thermal discharges can significantly affect the marine ecosystem; the reproduction, growth, and survival of larval and aquatic life forms. Regulatory agencies worldwide have therefore, set temperature limits for discharges from power plants to prevent the thermally induced destruction of aquatic species.[33] Fish, plankton and benthos are all affected in various degrees by thermal discharges from power plants.[34] The fear of the destruction of the marine eco-system and specifically, the impact on fish population because of thermal discharges has been the driving factor of opposition to the setting up of nuclear power plants at places such as Jaitapur and Kudankulam.

The process of thermal discharge is common to all major thermal and nuclear plants in India. To address concerns relating to the impact of thermal discharges on marine eco-system, the DAE under the aegis of various regional universities carried out thermo-ecological studies on India's inland as well as coastal NPPs.[35] These studies have shown that thermal discharges have a very localised as well as temporary impact on the aquatic species. The EIA report for the proposed NPP at Jaitapur, referring to the Kalpakkam study on marine ecology, claimed that 'no impact on marine life is anticipated from thermal discharges'.[36] The critics however rejected the conclusion in the EIA report on grounds that proposed project with the total capacity of 9900 MWe, is at least 25 times bigger than Kalpakkam NPP (440 MWe) and would discharge more than 6 million m3/day of warm water in to the sea.[37]

Finally, the public hearings for NPPs which are an integral part of the EIA process have been contested on various grounds. A common complaint against the public hearings conducted at places like Jaitapur and Gorakhpur, is the non-availability of detailed EIA reports prior to hearing.[38] Such instances in turn, prevent the participation of the affected people and inclusion of their concerns in decision making about the nuclear projects, thereby widening public distrust in the industry.

III) Nuclear Energy and Socio-Economic Concerns

Socio-economic concerns have been the primary drivers of many past and ongoing protests against nuclear power in India. The socio-economic concerns primarily arise from the conflict over natural resources like land, water and forests. The setting up of nucler plant requires sizable areas of land and huge quantities of water which often results in a conflict over these natural resources between local people and nuclear authorities. At many of the newly designated NPP sites, there have been serious conflicts over the natural resources between people and the government. The local people perceive the construction of nuclear installations as a threat to their livelihood.[39] Resolving the socio-economic concerns of local farmers and fishermen during the stages of site development and operations has been a serious challenge faced by the industry and respective state governments in India.

The agitation by the fishing community at the Kudankulam site became a politically sensitive issue after the members of local church sided with the protestors demanding closure of the power plant. The agitation was led by a local group called the People's Movement Against Nuclear Energy (PMNAE). The fisher folks are deeply concerned that the nuclear power plant would push them deeper into poverty and affect the food security of the entire southern Tamil Nadu and southern Kerala region.[40] The fishermen are opposed to one-kilometre security cordon around the plant which would be closed for fishing in order to ensure the physical safety of the plant.[41] They also fear that the low-grade waste discharge from the power plant would cause radioactive contamination of sea water and be hazardous to the marine life.[42]

Gorakhpur village in Haryana state has also witnessed protests for over a year against the four nuclear plants being set up in the region. The state government issued a notification to acquire land for the nuclear plant, mostly in the village of Gorakhpur in Fatehabad district.[43] The local people and residents of the other villages in the project area staged series of protests against the land acquisition.[44] In January 2012, these groups filed a public interest litigation in Punjab and Haryana High Court asking the court to quash the land acquisition notices and shift the project to barren or less fertile land in other districts.[45] During the public hearing held on July 17, 2012, farmers in Gorakhpur opposed the diversion of canal water for the nuclear plant as it would lead to serious scarcity of water for agricultural activities. India's land acquisition policy has traditionally been criticised for being insensitive towards the displaced population. It lacks the principle of participatory practices and merely focuses on compensation which is deemed seriously inadequate. The Supreme Court of India in responding to similar land acquisition related petition in 2011, ruled that private property of farmers should only be acquired if 'absolutely necessary'.[46] Further, the court ruled, "If land of such persons is acquired, not only the current but the future generations are ruined, and this is one of the reasons why the farmers who are deprived of their holdings commit suicide".[47] The Apex Court's ruling forced the states to reconsider their land acquisition policies in order to offer better rehabilitation and resettlement to the displaced people. The Haryana state government not only revised the earlier compensation package but also offered several incentives to re-settle the affected people, which were accepted by farmers of Gorakhpur, Baropal and Kajalheri villages.[48]

The Jaitapur site in Maharashtra has also been witnessing intense public opposition to the proposed nuclear power park in the region. The project is spread over 968 hectares of land, which encompasses five villages—Madban, Niveli, Karel, Mithgavane and Varliwada—which together have a population of 4,000. Madban and Varliwada have been identified for the site of the project proper, while Karel, Niveli and Mithgavane would host the township for the project staff.[49] The farmers complained that the government surveyors did not make a correct record of their plantation crops and also the compensation for the fruit trees is substantially less than what they earn from them annually.[50]

Besides farming and horticulture, the Jaitapur-Madban area has a large fishing community which is primarily opposed to setting up a nuclear park. The fishing community leaders fear that once the project becomes operational, its elaborate security arrangements would imperil fishermen's unhindered use of the two creeks of Jaitapur and Vijaydurg, where they get a depth of 20 fathoms that is usually found at a distance of two to three nautical miles on other coasts. Nate village, alone has 200 big trawlers and 250 small boats. Nearly 6,000 people are directly dependent on fishing in the area and more than 10,000 are dependent on ancillary activities.[51]

In Mithi-Vardi, Gujrat, the NPCIL is planning to set up six nuclear reactors designed by US-based Westinghouse, a subsidiary of Toshiba, Japan, with a total land requirement of 777 hectares. Lands of 340 farmers fall under the area earmarked for the proposed nuclear power plant. These farmers have submitted an affidavit to the district collector stating their unwillingness to give up agricultural land for setting up the nuclear power project.[52] Similarly, at the Kovvada site in Andhra Pradesh, people from surrounding villages have undertaken several hunger strikes against the land acquisition for the proposed nuclear park which will be set up by the NPCIL in collaboration with General Electric, USA.

The socio-economic concerns, as described above, have been the key driver of public opposition to mega industrial development projects in India. This has often witnessed a serious political turmoil in the form of prolonged stand-offs between government and the lay people. The public concerns over socio-economic issues together with nuclear safety and environmental concerns concert serious public opposition to nuclear energy. The perceived mismatch between the socio-economic risks and the benefits mainly shapes public attitudes to nuclear projects. Moreover, the involvement of local political groups in the public agitations makes it more difficult to arrive at mutually acceptable solution for all the stakeholders of socio-economic conflicts.

Public Concerns and Risk Management in India

The preceding sections have presented wide-ranging concerns expressed by people and civil society groups about nuclear power in India. Addressing these concerns is critical for the expansion of country's nuclear energy programme. The nuclear policy-makers often bemoan that risks associated with nuclear power projects are grossly exaggerated compared to the risks associated with other energy sources. Nevertheless, the DAE has initiated extensive risk management, communication and outreach efforts to neutralise fears about nuclear energy and address wide-ranging concerns of the people at various NPP sites. Presently, the risk management and communication efforts are being made by industry (the NPCIL), regulators (the AERB), and the government (the DAE) at both local and national levels. The risk management approach not only involves enforcing stringent safety measures but also requires bridging the information asymmetry and distrust that exists between scientific community and the lay people by highlighting various costs and benefits involved in nuclear power generation.

Post Fukushima, the debate over safety management of NPPs involved not merely the civil society members but also the highest political authorities in the country. The Indian Prime Minister issued directives to conduct an extensive safety review of the Indian NPPs to cope with extreme events as witnessed in Fukushima. Accordingly, the AERB and the NPCIL carried out two separate safety reviews of the operational NPPs in the country. The AERB formed a committee on March 19, 2011 to review the safety of Indian NPPs. The AERB task force examined the safety margins of the NPPs for 'Beyond Design Based Accidents' (BDBA), as well as the preparedness to respond to such events. The task force report concluded that the plants had sound designs and operating practices to cope with design basis as well as external events. It further noted that, "to add further margins and robustness, particularly to cope with beyond design-basis situations, additional safety upgrades have been recommended for implementation in a phased manner."[53]

The NPCIL too set up six task force groups to review the safety systems of Indian NPPs. The final report titled "Safety Evaluation of Indian NPPs Post Fukushima Incident" published in October 2011 concluded that there were no immediate safety weaknesses in India's NPPs and that the measures have been taken to prevent the accident like Fukushima. Additionally, the DAE also agreed to host the International Atomic Energy Agency's (IAEA) Operational Safety Review Team (OSART) for the safety peer reviews of Indian NPPs. The IAEA's OSRAT mission conducted the first safety inspections at Rajasthan Atomic Power Station (RAPS) during October 29, 2012 and November 14, 2012.[54]

In order to enforce effective regulatory management and usher in greater transparency in the regulatory mechanisms, the Indian government has decided to replace the current nuclear regulator—the Atomic Energy Regulatory Board (AERB), with the proposed Nuclear Safety Regulatory Authority (NSRA). The AERB plays an important role at all the stages; from plant siting to construction, commissioning and operation of the NPPs. However, the non-autonomous structure of AERB has time and again raised serious doubts about its institutional integrity.[55] Many non-governmental experts have questioned the AERB's ability to take stringent actions against nuclear operators on several occasions in the past. The incumbent AERB chairman sought to dispel some of the concerns about the AERB's autonomy, when he stated in a newspaper interview:

> It is a wrong notion that AERB is an extended arm of Department of Atomic Energy (DAE). AERB is independent of DAE while reporting to the Atomic Energy Commission (AEC). Reporting of AERB to AEC, in practical terms, comprise submission of Annual Report of AERB and the approval of the annual budget. AERB is functioning under the legislative framework of the Atomic Energy Act 1962 and the rules made there under; where adequate provisions have been made for regulation of safety. The existing arrangement has never come in the way of effective functioning of AERB as independent regulator. This is evidenced by the large number of enforcement actions taken by AERB over the years. In extreme situations, when found necessary, AERB has even ordered shutdown of plants or suspension of activities in projects.[56]

To enhance the legal status of the AERB, the government has introduced the 'Nuclear Safety Regulatory Authority Bill' (NSRA 2011l) in the Parliament. The creation of an independent NSRA shall restore the credibility of regulatory structure and practices in the public perception, while effectively separating the safety and promotional aspects of nuclear power in India. The government should therefore, expedite the creation of a new nuclear regulatory authority. The Indian government also needs to harness expertise in nuclear engineering outside the nuclear establishment to ensure independent nuclear regulation in the country. An independent regulatory institution will not only impart greater credibility to nuclear safety reviews of the NPPs but also be effective in communicating risks associated with nuclear installations.

The nuclear safety management also involves addressing risks to the environment and public health from low level radiation and the radioactive waste released from the NPPs. There has been considerable distrust relating to radioactive waste disposal between government and the public. The environmental concerns emerging from radioactive waste disposal have posed serious socio-political challenges for risks management, all over the world. In India, the DAE seeks to manage the problem of radioactive waste in the long run, by separating plutonium and other trans-uranic elements and by burning them in the fast neutron reactors. This will significantly lower the amount of waste material that would have to be stored in a final repository. While the programme for reducing the waste volume and its radioactive life span appears propitious, the DAE needs to communicate to the public, the costs and benefits involved in this strategy.

The low level radioactive releases from the NPPs into the surrounding environment are routinely monitored by the Environmental Survey

Laboratories (ESLs), set up at each nuclear site. The ESLs collect the baseline environmental data before commissioning of the NPP and thereafter, conduct systematic monitoring of various environmental matrices to analyse radioactivity levels during the operation of NPPs and Uranium mines. The environmental monitoring in the vicinity of NPPs for the last three decades has shown that radiation dosages among the members of the public are far below the AERB limit of 1000 mSv/y and negligible compared to the doses received from natural sources which is about 2400 mSv/y. Furthermore, the epidemiological studies conducted by the DAE have unambiguously concluded that there is no increased risk of various diseases, especially cancers, among people who live near nuclear facilities in India. Given the general lack of information on radiation releases from NPPs, the DAE has been promoting various communication and knowledge building efforts through Bhabha Atomic Research Centre (BARC) and various regional institutions. The DAE however, needs to allow greater consideration for environmental and public health values in its operational and regulatory practices. A greater transparency in the regulatory strategies and enhanced risk communication measures can significantly address some of environmental and public health related concerns.

Finally, the DAE has taken a series of initiatives which include offering local people a stake in the project; ensuring their incomes by providing assured direct/indirect employment; generating opportunities for them by providing the benefits of the project i.e. electricity and a share in profits through the life of the plant; and establishing local infrastructure for local villagers, etc. As part of the corporate social responsibility and the community engagement programme, the NPCIL in recent years has worked towards addressing local level needs and gaining trust of the people at the new NPP sites by undertaking local development projects and greater community partnership. The NPCIL, however, needs to enhance innovative public partnership programmes to improve community benefits of the new and existing NPPs in the country. Overall the government along with industry and regulators should adopt the best practices for managing risks from NPPs, and communicating it to those living near the existing or new nuclear facilities. The wave of global scepticism about nuclear energy after the Fukushima crisis has necessitated separating subjective risk perceptions against an objective assessment of risk, which can only be achieved through a coordinated risk communication efforts at local and national levels.

Conclusion

The nuclear energy programme in India currently faces unique public opposition which predominantly arises from various socio-economic, environmental and nuclear safety related concerns. Addressing the socio-economic concerns through improved governance measures can significantly minimise the opposition of local people at the new NPP sites in the country. Similarly, addressing environmental and nuclear safety related concerns require sustained communications efforts to inform the public about the improved safety measures enacted after the Fukushima accident. The increased risk communication efforts both at the regulatory and the operators' level can generate public trust and influence risk perceptions over the safety and environmental impacts of the nuclear installations. A well coordinated risk management and effective communication efforts are the best way to ensure public confidence and their wider participation in the nuclear energy projects.

NOTES & REFERENCES

1. 'Nuclear power plants pose unmanageable risks', *rsediff.com*, May 9, 2012, URL: http://www.rediff.com/news/slide-show/slide-show-1-nuclear-power-plants-pose-unmanageable-risks/20120509.htm, accessed on December 15, 2012. 'Nuclear safety: Apex court to hear PIL on Nov 14', *The Hindu Business Line*, November 11, 2011, URL: http://www.thehindubusinessline.com/industry-and-economy/economy/article2612059.ece, accessed on December 15, 2012, 'Indian anti-nuclear group demands moratorium on new reactor construction', *Environmental Justice,* URL: http://www.tni.org/article/indian-anti-nuclear-group-demands-moratorium-new-reactor-construction, accessed on December 15, 2012.
2. 'Petition in SC demanding stay on all N-plants', *Times of India,* October 14, 2011, http://www.timesnow.tv/Petition-in-SC-demanding-stay-on-all-N-plants/articleshow/4386459.cms, accessed on December 15, 2012.
3. Freudenburg, W.R. (1988), "Perceived Risk, Real Risk: Social Science and the Art of Probabilistic Risk Assessment", *Science,* 242 (4875), pp. 45-46; Rosa, E. and R.E. Dunlap (1994), "Nuclear Power: Three Decades of Public Opinion", *Public Opinion Quarterly,* 38 (1), pp. 295-301; and Slovic. P. (1987), "Perception of Risk", *Science,* 236 (4799), pp. 280-285.
4. Slovic, Paul (1987), "Perception of Risk", *Science*, 236 (4799), 280-285.
5. Ibid, pp.282.
6. Ibid, pp.283.
7. Costa-Font, J & C. Rudisill, E. Mossialos (2008), "Attitudes as an Expression of Knowledge and Political Anchoring: The Case of Nuclear Power in the United Kingdom", *Risk Analysis,* 28 (5), pp. 1273-1275.
8. Ibid, 1285-1287.
9. Roger E. Kasperson et.al. (1980), "Public Opposition to Nuclear Energy: Retrospect and Prospect, Science", *Technology, & Human Values*, 5 (31), pp.11-23.
10. Ibid, Costa-Font, et al. pp.1284-1285.
11. Whitfield, S.C., E.A. Rosa, A. Dan, and T. Dietz (2009), "The Future of Nuclear Power: Value Orientations and Risk Perception", *Risk Analysis*, 29(3), pp.425-437.

12. Ibid, pp.433-435.
13. Freudenburg, W.R. (2004), "Can We Learn from Failure? Examining US Experiences with Nuclear Repository Siting", *Journal of Risk Research*, 7(2), pp.153-166.
14. Slovic, P., J. Flynn, C.K. Mertz, M. Poumadere, C. Mays (2000), "Nuclear power and the public: A comparative study of risk perception in France and the United States." In O. Renn and R. Rohrmann (eds.), *Cross-cultural Risk Perception: A Survey of Empirical Studies*, Boston, MA: Kluwer Academic Publishers.
15. World Nuclear Association, "*Safety of Nuclear Power Reactors - updated 31 May 2012*", URL: http://www.world-nuclear.org/info/Safety-and-Security/Safety-of-Plants/Safety-of-Nuclear-Power-Reactors/#.UaHqK4k3Ciw, accessed on January 15, 2013.
16. Ramanna, M.V. (2011), "Beyond our imagination: Fukushima and the problem of assessing risk", *Bulletin of Atomic Scientists*, April 19, 2011, URL: http://thebulletin.org/web-edition/features/beyond-our-imagination-fukushima-and-the-problem-of-assessing-risk, accessed on January 17, 2013.
17. The Chernobyl accident offered many important lessons for radiation and environmental safety. According to the United Nations Scientific Committee on the Effects of Atomic Radiation (UNSCEAR) report on Chernobyl accident, "among the residents of Belarus, the Russian Federation and Ukraine, there had been up to the year 2005 more than 6,000 cases of thyroid cancer reported in children and adolescents who were exposed at the time of the accident". The report notes that more cases of thyroid cancer can be expected during the next decades. Apart from instances of thyroid cancer, the report states that, "there is no evidence of a major public health impact attributable to radiation exposure two decades after the accident. The incidence of leukaemia in the general population, one of the main concerns owing to the shorter time expected between exposure and its occurrence compared with solid cancers, does not appear to be elevated." Many independent scientists and researchers, however, do not agree with the UNSCEAR findings and argue that there is an increased rate of Leukaemia among children and rescue workers in Chernobyl. In 2003, to settle the prolonged controversy surrounding the impact of Chernobyl disaster, the IAEA and other international organizations convened the Chernobyl Forum comprising experts from various scientific streams. In September 2005, the Forum concluded that "the total number of people that could have died or could die in the future due to Chernobyl originated exposure over the lifetime of emergency workers and residents of most contaminated areas is estimated to be around 4000". The definite impact of radiation on human health therefore continues to be debated by different scientific groups. See, Ivanov V K. (2007), "Late cancer and non-cancer risks among Chernobyl emergency workers of Russia", *Health Physics*, Vol. 93, pp. 470–79, Noshchenko A G, Moysich K B, Bondar A, Zamostyan P V, Drosdova V D, Michalek A. M. (2001), "Patterns of acute leukaemia occurrence among children in the Chernobyl region" *International Journal of Epidemiology*, Vol. 30, pp. 125-29, and Williams D, Baverstock K. (2006) "Too soon for a final diagnosis", *Nature*, 440:993–94.
18. Gadekar, Sanghamitra and Gadekar, Surendra (2011), "*The Real Costs of Nuclear Power*", People's Tribunal on the Safety, Viability and Cost Efficiency of Nuclear Energy. Dr Gadekar in his study registered around one hundred cases of congenital deformities in villages surrounding Jaduguda. Another study at Rawatbhata nuclear plant site shows the higher rate of miscarriages, still births and one day deaths of new born babies as well as the rise in cases of solid tumours.
19. Rahman, Shakeel ur, "Study on Health Status Of Indigenous People Around Jadugoda uranium mines in India", *Indian Doctors for Peace and Development*, URL: http://ippnw.org/pdf/jadugoda-health-survey.pdf, accessed on January 16, 2014.
20. Parthasarathy, K.S. (2008), "Use uranium in nuke power", *PTI*, March 31, 2008.
21. Nuclear Power Corporation of India Limited Press Release, "*No Health Hazard from Nuclear*

Power Plant Operation", June 11, 2011, URL: http://www.npcil.nic.in/pdf/No_Health_Hazard_from_NPP_Operation.pdf, accessed on January 20, 2013.

22. Ibid, See, Pramesh, C.S. and Badwe Rajendra (2012), "Nuclear Radiation and Health—the True Story", *Nuclear India,* Vol. 47 (1), Jan-Mar 2012, p. 47. During January and December 2006, the Health Physics Unit of Bhabha Atomic Research Centre (BARC) carried out a health survey of 598 villagers residing near Jaduguda which included 152 males, 217 females and 229 children below the age of twelve. As per the survey results in this group, the villages around the uranium mines, mills and tailings ponds have been surveyed from time to time by agencies appointed by government and the reports have conclusively suggested that the incidence of diseases including cancer in these areas is not higher to that of national average. The survey specifically mentions that there is no detected case of congenital malformation and mental retardation; no cancer cases were found; the number of infertility cases amongst married woman was two which is below the national average; and the villagers suffer from conventional health problems which could be seen in any village with similar socio-economic conditions. A study conducted by the Tata Memorial Hospital in Tarapur (Maharashtra) and Kakrapar (Gujrat) involved over 10,000 individuals working at the nuclear power plants, their families and children. The study has shown that no increase in cancer incidence was witnessed amongst all the categories compared to those who were not exposed to radiation. There was no increase in the incidence of birth defects among children born to radiation workers. The Kakrapar study showed no cancer among those living in the area. The Arinagar Anna Memorial Cancer Institute conducted similar study in Kalpakkam involving more than 5000 radiation workers, nearly 4000 spouses and over 5000 children born of these workers. No differences were seen in cancer incidence between radiation and non-radiation workers; there was no difference in cancer incidence between radiation workers at Kalpakkam and non-radiation workers at BARC, Mumbai. Similarly, studies conducted at the Rawatbhata atomic power station, by SMS Medical College, Jaipur and at the Narora atomic power station by Jawaharlal Nehru Medical College came to similar conclusion as that of the previous studies.
23. A Gopalakrishnan (2011), "Why should Jaitapur be made a guinea pig for untested reactor?", *Daily News Analysis (DNA)*, March 17, 2011, URL: http://www.dnaindia.com/mumbai/comment-why-should-jaitapur-be-made-a-guinea-pig-for-untested-reactor-1520843, accessed on January 20, 2013. Dr Gopalakrishan has contended that the reactor has to be physically built and then only it can be tested, and the EPR is therefore a totally untested reactor, even if Areva claims they have combined various best design features on paper in conceiving the EPR. The reliability and safety of EPR will be extremely low and unknown until, through different stages of operation and testing over years, all indicated problems are rectified.
24. The central government set up 15 member expert committee for the KKNPP which interacted with officials of the state government of Tamil Nadu and spokespersons of the people in the neighbourhood of Kudankulam project. See the report, *Supplementary Report on Safety of KKNPP and Impact of Its Operations on Surroundings,* Expert Group Constituted by GOI, January 31, 2012.
25. A Gopalakrishnan (2012), "Stop Kudankulam fuelling, lives are at stake!", *Daily News Analysis (DNA)*, September 24, 2012, URL: http://www.dnaindia.com/analysis/column-stop-kudankulam-fuelling-lives-are-at-stake-1744417, accessed on January 20, 2013.
26. Many eminent thinkers have strongly pitched in favour of nuclear energy as a solution to GHG reduction, See Brook, B.W. & Lowe, I. (2010), *Why vs Why: Nuclear Power*, Sydney: Pantera Press; Christine Todd Whitman (2011), "Nuclear power needs to remain central to our energy mix", *The Hill*, November 15, 2011, URL: http://thehill.com/opinion/op-ed/193835-nuclear-power-needs-to-remain-central-to-our-energy-mix#ixzz2WwVnTX7N, accessed on January 21, 2013. George Monbiot (2011), "Why Fukushima made me stop

worrying and love nuclear power", *The Guardian*, March 21, 2011; Gwyneth Cravens (2007); "*Power to Save the World: The Truth About Nuclear Energy*, Vintage Press; James Lovelock (2000), "*Gaia: A new look at life on earth*", New York: Oxford University Press.

27. Gadgil, Madhav (2013), "Ratnagiri and Sindhudurg districts: Summary draft report of the Maharashtra Government consultation" *Western Ghats Ecology Expert Panel*, URL: http://www.dianuke.org/wp-content/uploads/2011/04/ratnagiri-sindhudurg-environmental-study-tour-report_madhav-gadgil-committee.pdf, accessed on January 20, 2013.
28. Ministry of Environment and Forests (1994), "Environment Impact Assessment Notification S.O.60(E) - dated 27/01/1994", New Delhi, URL: http://envfor.nic.in/sites/default/files/so-60(e).pdf, accessed on January 22, 2013
29. Divya Badami Rao, M.V. Ramana (2008), "Violating Letter and Spirit: Environmental Clearances for Koodankulam Reactors", *Economic & Political Weekly*, December 20, 2008.
30. Ibid, Divya Badami Rao, M.V. Ramana, p.15.
31. People's Movement Against Nuclear Energy (2011), "*Thirteen Reasons Why We Do Not Want the Koodankulam Nuclear Power Project*", August 25, 2011, URL: http://www.dianuke.org/thirteen-reasons-against-the-koodankulam-nuclear-power-project/, accessed on January 24, 2013
32. Rao, K.R. (2001), "Radioactive waste: The problem and its Management", *Current Science*, Vol. 81 (12), December 25, 2001, pp. 1534-1546; Friends of the Earth Australia, *Reprocessing*, URL: http://www.foe.org.au/anti-nuclear/issues/nfc/waste/reprocessing, accessed on January 25, 2013; Greenpeace - Media Briefing, "Nuclear Reprocessing at Sellafield", November 2001, URL: http://www.greenpeace.org.uk/MultimediaFiles/Live/FullReport/4372.pdf, accessed on January 25, 2013; Union of Concerned Scientists, "Nuclear Reprocessing: Dangerous, Dirty, and Expensive", May 04, 2011, URL: http://www.ucsusa.org/nuclear_power/nuclear_power_risk/nuclear_proliferation_and_terrorism/nuclear-reprocessing.html, accessed on January 25, 2013.
33. Paschoa, A.S. (2004), "Environmental Effects of Nuclear Power Generation", in Jose Goldemberg (Eds.), *Interactions: Energy/Environment in Encyclopaedia of Life Support Systems (EOLSS)*, UNESCO, Eolss Publishers, Oxford, UK.
34. Ibid, Paschoa, A.S.
35. Apte, Shree Kumar (2002), "The DAE-BRNS initiative on thermal ecological studies relevance and implications", *Bhabha Atomic Research Centre*, Mumbai, July 01, 2002.
36. National Environmental Engineering Research Institute (2010), "Executive Summery - Environmental Impact Assessment for Proposed Jaitapur Nuclear Power Park - Volume I: EIA (Main Report)", September 2010, p. S-34-35.
37. Centre for Science and Environment (2010), *Technical Analysis of EIA report of 9900 MW Nuclear Power Plant (NPP) at Jaitapur*, URL: http://cseindia.org/node/1553, accessed on January 25, 2013.
38 Jaitapur Nuclear Power Plant Public Hearing Transcript in Marathi, *Lokayat*, April 16, 2010, URL: http://lokayatpune.wordpress.com/2010/07/05/jaitapur-nuclear-power-plant-public-hearing-transcript/, accessed on January 25, 2013.
39. M. Sprangler (1974), "Environmental and Social Issues of Site Choice for Nuclear Power Plants", *Energy Policy*, 2(18), pp. 18-19.
40. Latha Jishnu, Arnab Pratim Dutta, Ankur Paliwal (2012), "Kudankulam meltdown", *Down to Earth*, April 15, 2012, URL: http://www.downtoearth.org.in/content/kudankulam-meltdown, accessed on January 27, 2013
41. Ibid, Latha Jishnu et al.
42. Ibid, Latha Jishnu et al.
43. Executive Summary of Environmental Impact Assessment Report - 4 x 700 MW PHWR Gorakhpur Haryana Anu Vidyut Pariyojana (GHAVP), URL: http://www.npcil.nic.in/pdf/

Executive_Summary_of_Draft_EIA_of_Gorakhpur_Project_English.pdf, accessed on January 28, 2013.

44. Sushil Manav (2012), "Gorakhpur nuclear power plant: Three company officials held captive", *Tribune News Service—Fatehabad*, February 6, 2012, URL: http://www.tribuneindia.com/2012/20120207/haryana.htm#3, accessed on January 29, 2013.
45. Jyotika Sood (2012), "Farmers move court against Gorakhpur nuclear power plant in Haryana", *Down to Earth*, January 25, 2012, URL: http://www.downtoearth.org.in/content/farmers-move-court-against-gorakhpur-nuclear-power-plant-haryana, accessed on February 02, 2013.
46. Supreme Court of India: Civil Appellate Jurisdiction: Petition for Special Leave to Appeal (CIVIL) No. 28034/2011, Petitioner: State of Haryana Versus Respondents: Mukesh Kumar & Others, URL: http://www.sci.nic.in/outtoday/38720.pdf, accessed on February 02, 2013.
47. Ibid.
48. Bhatia, Varinder (2012), "Haryana n-plant project clears land hurdle", *Indian Express*, September 24, 2012.
49. "Jaitapur nuke plant will be a social disaster: TISS report", *DNA Mumbai*, Dec 26, 2010, URL: http://digital.dnaindia.com/epaperpdf%5C27122010%5C26main%20edition-pg3-0.pdf, accessed on February 05, 2013
50. Mahesh Kamble (2010), "People's Report on Social Impact Assessment of Jaitapur-Madban NPP", *Tata Institute of Social Sciences*, August 15, 2010. The author notes that survey officers evaluation per tree of mangoes is Rs. 9,386, whereas they earn Rs 10,000-15,000 from a single tree annually; and it is Rs 1,989 for a plant of cashew whereas the annual earning per plant is usually Rs 4,000-5,000.
51. Anusha Subramanian (2011), "Cold war at Jaitapur: Villagers worry over mode of living", *Business Today*, April 03, 2011, URL: http://businesstoday.intoday.in/story/jaitapur-villagers-protest-against-nuclear-power-project/1/13890.html, accessed on February 07, 2013. Muffazal Phadnis, a fisherman in Jaitapur village points out that over 6,000 people in these villages depend directly on fishing for a living. We do not own agricultural land. Secondly, Majith Govalkar, a boat owner in Jaitapur points out that, "Our understanding is that the reactors will discharge hot water into the sea, upsetting the ecology and scaring away the fish. Given the security concerns of a nuclear plant, the government is also sure to create a security zone and ban fishing near the coast".
52. Adam Halliday and Hiral Dave (2012), "Gujarat n-plant runs into land acquisition hurdle", *Indian Express*, August 21 2012, URL: http://www.indianexpress.com/news/gujarat-nplant-runs-into-land-acquisition-hurdle/990805/0, accessed on February 12, 2013. A survey of fruit trees conducted in Mithivirdi and Jasapara villages accounted for more than 10,000 mango trees, 1,000 chikoo and custard-apple trees each in both the villages.
53. See Recommendations and Conclusions, "Report of the AERB Committee to Review Safety of Indian Nuclear Power Plants Against External Events of natural Origin", August 2011, ULR: http://www.aerb.gov.in/AERBPortal/pages/English/t/publications/CODESGUIDES/report-nov.pdf.
54. IAEA Press Release 2012/27, "IAEA Leads Operational Safety Mission to Rajasthan Atomic Power Station 3 and 4", *IAEA*, November 14, 2012, URL: http://www.iaea.org/newscenter/pressreleases/2012/prn201227.html, accessed on March 05, 2013.
55. Purkayastha, Prabir (2012), "A Toothless Nuclear Safety Authority", *Delhi Science Forum*, January 17, 2012, URL: http://www.delhiscienceforum.net/peace-and-disarmament/450-a-toothless-nuclear-safety-authority.html, accessed on April 22, 2013. Also, the performance audit report of the AERB issued by Comptroller and Auditor General of India (CAG) published in August 2011 have severely criticised the regulators on numerous grounds. The CAG criticised the AERB for not preparing nuclear safety policy despite having a mandate

to do so since 1983; failing to prepare 27 of 168 safety documents; not having a detailed inventory of all radiation sources; and failure to adopt international practices. The CAG report was widely referred by the critics who questioned the credibility of the AERB in terms of maintaining the regulatory professionalism.

56. AERB' Chairman's Interview, "AERB is independent, not DAE's extended arm: SS Bajaj", *Business Standard*, April 27, 2013, URL: http://www.business-standard.com/article/economy-policy/aerb-is-independent-not-dae-s-extended-arm-ss-bajaj-113042700328_1.html, accessed on February 15, 2015.

CHAPTER 4

Safety of India's Nuclear Power Plants

Sudhinder Thakur

Introduction

The March 2011 Great East Japan Earthquake and tsunami caused extraordinary damage to the Fukushima Dai-ichi nuclear plant because of severe flooding, leading to large-scale radiological fallout. The first generation BWR reactors at Fukushima sustained serious core degradation as well as simultaneous multiple unit failures due to the catastrophic external events. The unprecedented disaster at Fukushima exposed the vulnerabilities of NPPs to extreme external events. While, the Fukushima accident reveals no fundamental flaw in nuclear technology, it did call for pertinent safety improvements from the new lessons learned. The protection of public health and safety from radiological releases has been, and continues to be, the primary focus of nuclear reactor safety. A salient issue arising from the accident at the Fukushima Dai-ichi Plant is therefore the need for ensuring adequate protection from catastrophic external events known as 'beyond-design-basis events' (BDBE). The BDBE events are also termed as events, also termed as low-probability, high consequence events.

Post Fukushima accident, the Atomic Energy Regulatory Board (AERB) has undertaken exhaustive safety reviews; analysed important lessons; and made recommendations for improving safety at Indian nuclear power plants. The Fukushima crisis clearly highlighted that the severe accident management infrastructure is extremely important, and it will become even more important as an all-risk approach is incorporated into the nuclear safety paradigm. This paper, first reviews the reactor safety management systems in India's nuclear power plants (NPPs) to ensure adequate protection. It discusses the defence-in-depth principle and additional capabilities under consideration to extend the design to cover severe accident response. The paper also discusses the various measures undertaken to cope with conditions of station blackout, and others to ensure the availability of the ultimate heat sink etc.

Philosophy of Nuclear Power Generation

Conceptually, a nuclear plant is like a coal fired power plant; the significant difference being the way steam is produced. In a fossil power plant, coal/gas is burnt to convert water into steam. This steam is then sent to a turbine, where heat energy is converted into mechanical energy. The coupled generator converts mechanical energy into electrical energy which is transmitted to the end users for consumption. In a nuclear power plant, steam is produced from heat of fission of atoms. Steam once produced is used to drive the turbine-generator, much like the thermal plants, to produce electricity. Figure 1 shows the similarities and the differences between the nuclear and coal thermal power stations.

Figure 4.1: Thermal and Nuclear Power Plant

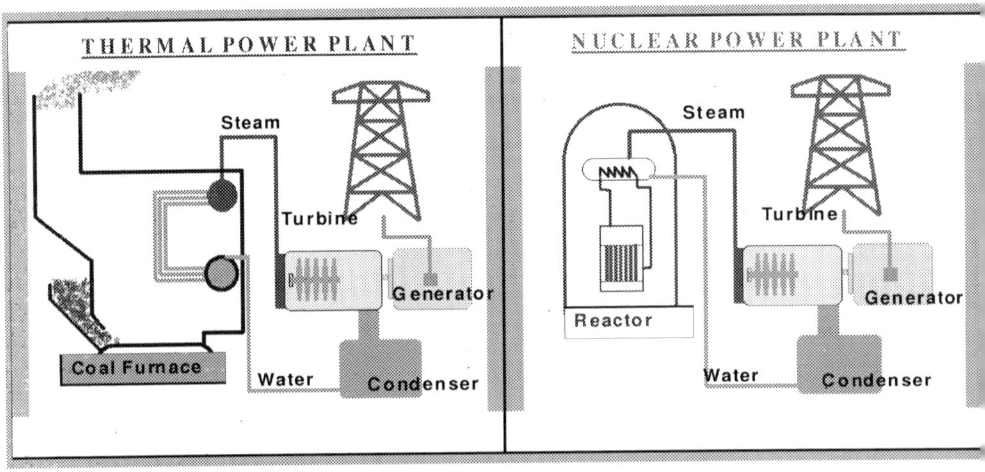

In a nuclear plant, the heat for production of steam comes from fission of uranium atoms. The typical equation for fission of uranium is;

$$_{92}U^{235} + {_0}n^1 \rightarrow {_{56}}Ba^{142} + {_{30}}Kr^{91} + 3{_0}n^1$$

When a uranium atom of atomic no 92 and mass number 235 (present as 0.7% in natural uranium, most of the rest being uranium 238) encounters a neutron under specified conditions, it undergoes fission; producing two fission products, BA & Kr and also three neutrons. An accurate measurement of masses of the starting constituents on the left of the equation with those on the right side reveals a missing mass. This missing mass appears as energy in accordance with the equation:

$$E = mC^2$$

Where E: Energy produced
m: Missing mass
C: Velocity of light

The fission of one uranium atom gives 200 million electron volt of energy as compared to about 13 electron-volts from combustion of one atom of carbon. The nuclear fuel is thus a very intense source of energy and very small quantities, relative to coal, of fuel are required for production of a given quantity of electricity.

Figure 4.2: Fission of Uranium 235

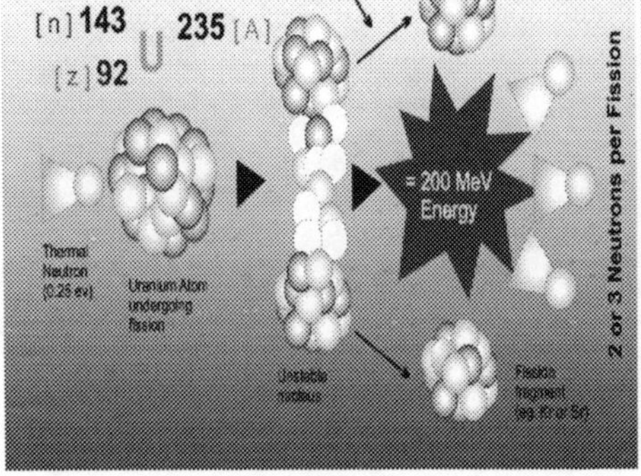

Figure 4.3: Energy Release-Coal Combustion vs. Uranium Fission

A. Cola Combustion

$_6C^{12} + O_2 \rightarrow CO_2$ + Energy (12.1 eV)

B. Nuclear Fission

$_{92}C^{235} + {_0n^1} \rightarrow X + Y$ + Energy (200×10^6 eV)

Energy released by fission of 1 atom of U is million times more than the energy released by combustion of 1 atom of Carbon

The high energy intensity of nuclear fuel implies a significant reduction in fuel transportation infrastructure along with reduced losses and risks associated with fuel transportation. The nuclear fuel transportation costs are negligible and therefore not impacted by the escalation and are practically distance neutral from the mines. Another advantage of the high energy intensity is the storage of fuel. Fuel for years of operation of a power reactor can be safely stored at site in a small storage facility and in terms of security of supply offers an unparalleled advantage over other fuel options like coal or gas.

Working of a Nuclear Power Plant

The primary function of a nuclear reactor is the heat production using radioactive fuel. Nuclear plants use uranium oxide as fuel. Uranium oxide, in the form of small (one cm diameter and one cm length) high density (10.5grams/cc) pellets is encapsulated in a metallic tube which is sealed at both ends. While the heat produced in these uranium oxide pellets during fission is transferred to the coolant which passes over the tubes, the products of fission remain within the encapsulated tubes. Thus the fuel externally, after the energy has been extracted, physically, looks the same as a fresh fuel bundle. The coolant does not come into physical contact with uranium pellets.

Figure 4.4: Fuel Pellet

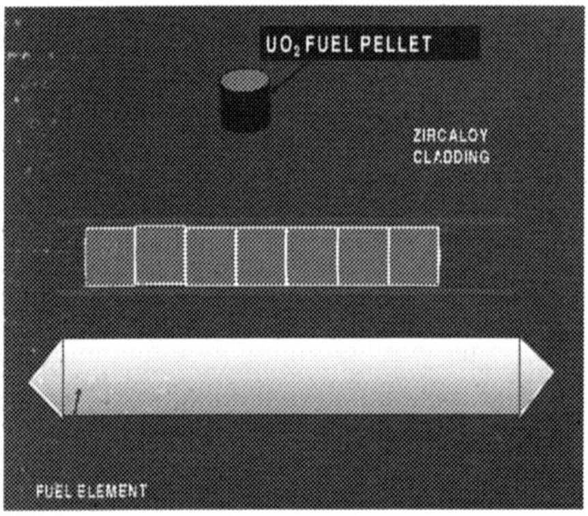

Figure 4.5: Fuel Bundles for Nuclear Power Reactors

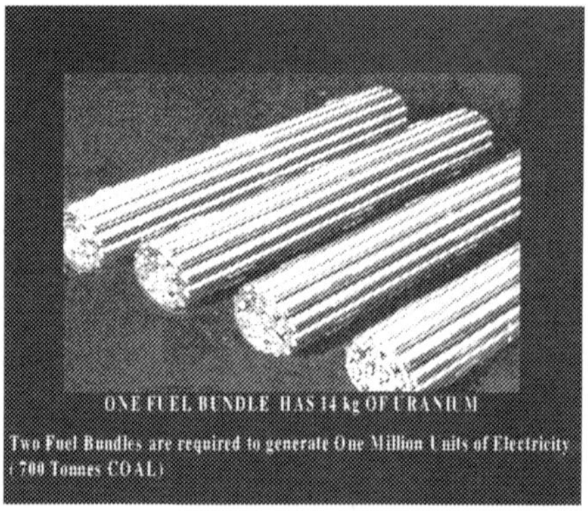

The fission equation $({}_{92}U^{235}+{}_{0}n^{1}\rightarrow{}_{56}Ba^{142}+{}_{30}Kr^{91}+3{}_{0}n^{1})$ also produces three neutrons which after losing some energy, can cause fission of more atoms and thus used for producing further fission. Thus a chain reaction and continuous heat production is achieved. Since three neutrons are produced for fission of one atom and only one neutron is required for sustaining chain reaction, excess neutrons are available after considering leakages and absorption. This neutron interacts with uranium 238 (abundant constituent of natural uranium) and produces plutonium which is again a fissile material after chemical separation and reprocessing of the used fuel. Thus all nuclear power reactors while producing power also produce additional fuel.

The reactor, apart from containing the uranium fuel, also has control rods to regulate the operation of the reactor including automatic shutdown of the reactor (i.e. stopping nuclear fission process). This heat produced by the fission process is extracted by circulating coolant over the fuel bundle. The coolant does not come in contact with the fuel. The heat extracted from the reactor is exchanged in a steam generator where water in another loop is converted into steam. The steam is fed to the turbo-generator to produce electricity.

The typical fission equation ($_{92}U^{235}+_{0}n^{1}\rightarrow_{56}Ba^{142}+_{30}Kr^{91}+3_{0}n^{1}$) also yields two fission products. There are a large number of fission equations and accordingly some 200 fission products are produced in a reactor. Most of these fission products undergo transformation till a stable products stage is reached. Radiation is also produced in these processes and nuclear safety is directed at containing these radiations from escaping to the environment.

The radioactive fission products are generated and contained within the solid pellets of uranium-dioxide fuel. In addition these fuel pellets are inside hermetically sealed zirconium alloy tube. Providing adequate cooling at all times, including when the reactor is shut down, ensures integrity of pellets and the zirconium tubes thereby containing the radioactive products within these tubes. The cooling of fuel in proportion to the heat generated at all times is therefore an important target for ensuring safety. Nuclear power plants during normal operation, add practically nothing to background radiation. Proper safety systems in sitting and design are incorporated to avoid accidents and to contain the consequences, in the high unlikelihood, of their happening.

Safety Significance of Nuclear Plants

The fission process that produces heat also produces radiation and other fission products. The fission products, though contained in the encapsulated tubes, continue to emit radiation and heat even after the initial fission process is terminated. The entire gamut of nuclear safety is directed to ensure that the workers, public and environment are not exposed to any risks from nuclear radiation. Radiation is a normal part of everyday life and it is present all around us, everybody is exposed to the radiation.

Figure 4.6: Schematic Diagram of Indian PHWR

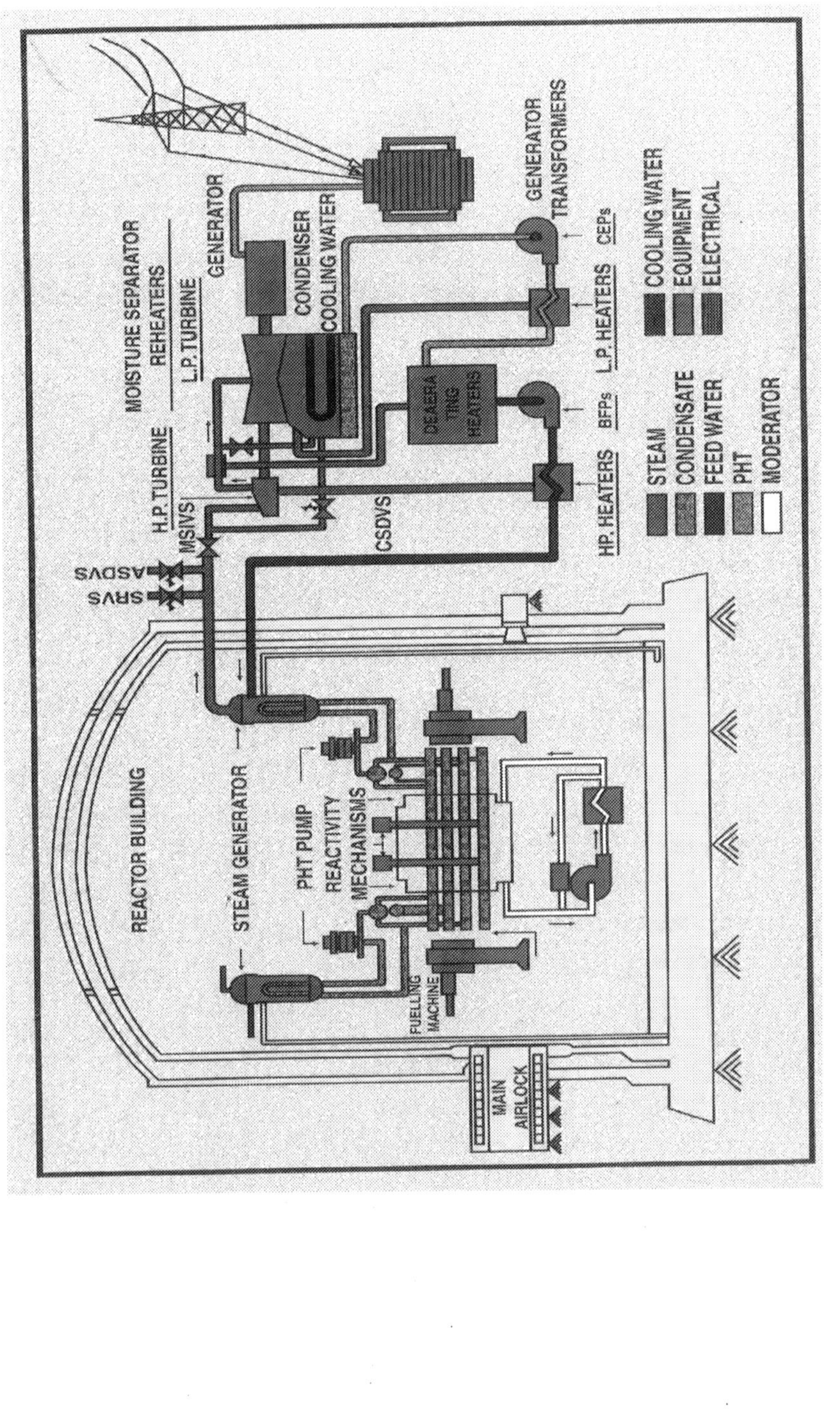

Figure 4.7: Radiation

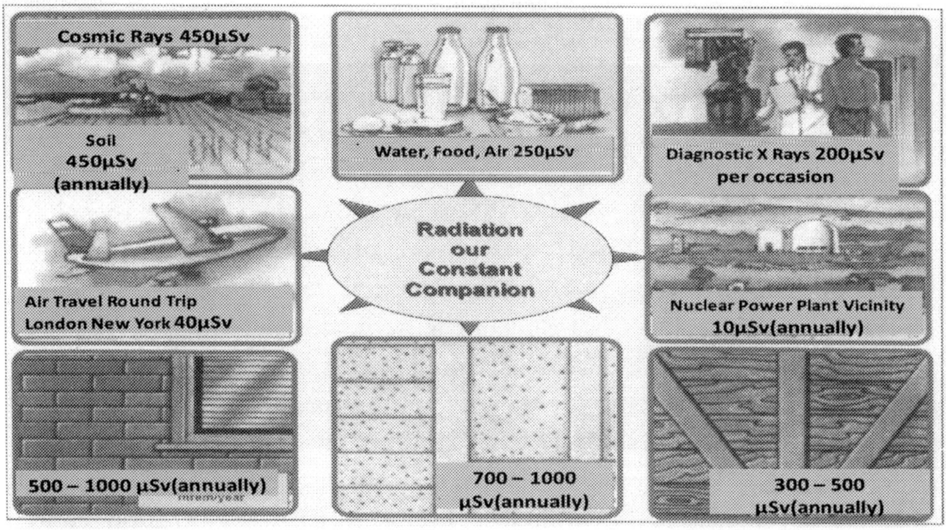

On an average, every individual receives 2400 micro sievert of radiation in a year from natural sources. The background radiation is measured by different Indian Environment Radiation monitoring Network (IERMON) stations. It can be seen that the background radiation is widespread and is not linked to presence or otherwise of nuclear installations.

Additionally, one gets 50 Micro-Sieverts (μSv) in the course of one chest X-ray; or 40 (μSv) from an 8 hour round air trip. The maximum limits imposed for an astronaut per mission is 250,000 μSv. Similarly the maximum limit imposed on nuclear occupational workers is 20,000 μSv per year. The exposure of the public from nuclear operations is much lower and a very conservative limit of 1000 μSv per year is specified. This level of exposure is one thousandth of the levels of some temporary effects of radiation. In addition the actual exposure from nuclear facilities is only a fraction of these limits and is well within variation of the natural background radiation. The stress on nuclear safety is explicit from the importance safety gets in all phases of nuclear power development; siting, design, construction and operation.

Safety in the Siting of NPPs

The location of nuclear power plants, in the first instance itself is finalised after detailed consideration for ensuring safety. Investigations are undertaken to ensure that the location of the plant at a given site will not pose undue radiological load to the public and the environment. The effects of both natural and man-induced external events on the plant are assessed to ensure that they will not interact with the plant, in a way, that could lead to unacceptable radiological consequences. Site-related natural external events form the design basis of the plant and include seismic activity, flooding and extreme meteorological events. NPP sites in India are so selected that sites with man-made external events such as: air crashes, chemical explosions and toxic gas release, blasting and mining operations, drilling etc in the vicinity of the plant are excluded. The detailed criteria are given in the regulatory code of practice for site evaluation. Geography, demography and topography of area around site, land use, land cover, food/milk production, dietary habits, adjoining industries chemical plants, military installations, gas pipelines, airports, archaeological monuments, parks and places of pilgrimage, including anticipated changes in their utilisation are included in site evaluation report (SER). The SER also includes site characteristics affecting safety and interactions of NPP with its environment.

Data about meteorological conditions, frequency of the occurrence and possible consequences of extreme meteorological conditions such as tornadoes, hurricanes or typhoons, cyclones and precipitation, distribution of wind velocity and direction, atmospheric diffusion and transport, cooling water requirement and sources, ground water movement, dispersion conditions, potable and service water supplies are collected and used as an input for the design. The effects of natural phenomena as tidal effects, floods, coastal cyclones, tsunamis etc consequences of failure of upstream and downstream dams are studied in detail. The design basis flood level and safe grade elevation of the site are worked out, based on the above information.

A region within a radial distance of 300km from the site is studied for collection of relevant seismic data. Data and information from geological, geophysical, seismological and geotechnical investigations are collected and analysed. All faults in the region are investigated to determine their capabilities. A map of lineaments is prepared on the basis of Landsat imageries, aerial photographs, geological, geo-morphological, geo-physical and micro-seismic data and examined carefully along with historical earthquake data for identification of any active fault. Epicentres of all earthquakes (Magnitude 3.0) are superimposed on the lineament map. The magnitude or epicentral intensity of earthquakes, the depth of focus and distance from the site are evaluated. A Design Basis Earthquake (DBE) is established to derive the Peak Ground Acceleration for each postulated earthquake event that all equipment & structures must withstand. A nuclear plant is designed to withstand the effects of vibratory ground motion arising from earthquakes. The design basis ground motion (DBGM), for this purpose, is evaluated for each site. The DBGM is derived for two levels of earthquakes: the severe earthquake level is termed as S1 or Operating Basis Earthquake (OBE) level; and the extreme level is designated as S2 or Safe Shutdown Earthquake (SSE) level. The OBE, (with Mean Recurrence Interval MRI ~100 years) is reasonably expected to be experienced at the site area once during the operating life of the plant. All plant systems are designed to continue to function when subjected to OBE. The SSE is the maximum potential vibratory ground motion that can be expected in the region (with Mean Recurrence Interval (MRI) > 10,000 years). It represents the maximum level of ground motion to be used for design of safety related structures, systems, components and equipment of NPP.

Safety Principles and Engineered Safety Features

Defence-in-depth consists of well-defined principles that are incorporated in the design of the nuclear power plant. The various levels of this defence-in-depth are:

First Level

All the components and systems are designed using conservative safety standards. The nuclear power plant design is in conformity with international codes and standards. This ensures that best global practices are followed and that there are sufficient margins in design and that the fabrication, construction and inspection for systems provide high quality throughout. The systems so designed are expected to function with sufficient assurance that the systems have the inherent capacity to withstand deviations from normal operation.

Second Level

The design includes systems to detect, intercept and control any deviations in normal operation. In case the deviations are greater than the scope of control systems, automatic protective systems are provided to prevent the deviations leading to the accidents. One such protective feature is provision of the fast automatic shutdown of the reactor, in the event of any of the selected plant parameters exceeding the pre-set limit. This feature provides a high degree of assurance that prompt shutdown of the reactor will be achieved to curb plant disturbances. In the current Indian pressurised heavy water designs two automatic fast acting and independent shutdown systems of diverse design have been included, although one system is adequate for handling any situation.

Third Level

As a matter of abundant caution, additional safety features are provided to ensure that the consequences of accidents are mitigated. The means to ensure continued cooling of fuel even under accident conditions and containing radioactivity are provided so that radioactivity is not released into the environment. The major safety systems in this category are:

- Emergency core cooling system.
- Containment system.

Fuel cooling during the normal operation of the plant is ensured by the transfer of generated heat to produce steam and run the turbine. During reactor shut down the small amount of "decay heat" (about 2 per cent of the full power, couple of minutes after shut down, that gradually decreases to about 0.5 per cent after 24 hours) can be eliminated quite easily as long as the primary boundary is intact. However, special provisions are made in the design to ensure that adequate cooling of the fuel takes place, even in an unlikely event of a rupture in the primary cooling system. In such an event condition, there is provision for injecting water through an emergency core cooling system (ECCS), so as to maintain the cooling of the nuclear fuel. Fuel design provisions hinder the overheating of fuel, preventing escape of radioactivity.

An added level of safety is provided by the containment building, surrounding the reactor. It is a leak-tight barrier around the reactor, designed to control the escape of radioactivity into the environment even in the event of large releases of radioactivity from the fuel. A salient feature of the design is the double containment principle. In double containment, even if there were to be some release of radioactivity from the inner building (primary containment), most of it is trapped in the annular space between the inner and outer buildings (secondary containment), which is maintained at negative pressure.

Fourth Level

Finally, as a matter of safety policy in India, an exclusion area is maintained around the station, where no public habitation is allowed. Thus the defence-in-depth approach provides for a conservative design with control and protective features to prevent occurrence of an accident. However, if one should occur at all, there are systems to confine radioactivity by engineered mitigating safety features. In physical terms, there are series of barriers in a nuclear reactor between radioactive fission products in a solid fuel matrix and the public. Escape of radioactivity from the fuel will therefore require a breach in four barriers, namely: Fuel, Fuel Cladding, Heat Transport System and Containment Building. The figure below depicts the different barriers to the release of radioactivity to the environment.

Figure 4.8: Barriers to Release of Radioactivity

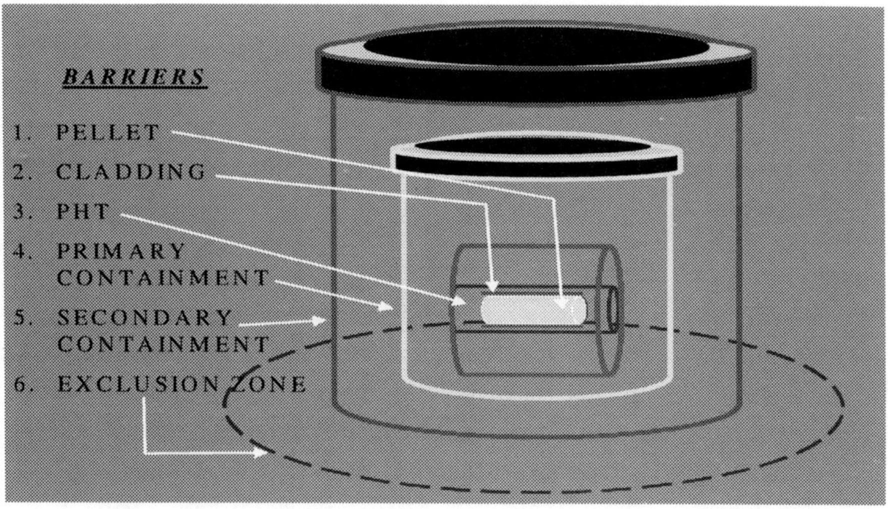

It is thus clear that the defence-in-depth approach involves physical barriers between the radioactivity and environment. These barriers (fuel matrix itself, zircaloy tube over fuel pellets, the high pressure boundary, double containment buildings and the separation between the reactor and the public) ensure that there is no impact on environment outside the reactor. Over and above these provisions, there are procedures for managing any radiation emergencies, and these are covered under "Emergency Preparedness" for plant, site and off-site emergencies. The procedures relating to the organisation, responsibility and operations in the case of "plant" and "site" emergencies are prepared by concerned plant, reviewed and approved by AERB. The documents in case of off-site emergencies are approved by the concerned state Government. In case of an "off-site" emergency, the district collector is the "off-site Emergency Director". Periodic exercises for all three types of emergencies are conducted as per procedures at all power stations. These mock exercises ensure preparedness at all times and ensure that emergency preparedness is strengthened by correcting any deficiencies observed during these exercises.

Design Principles for Safety Systems

Physical and functional separation is built into the design for all systems, components and structures important to safety. The physical and functional separation of systems/components ensures that a single local event (fire, missile, pipe failure) will not impair the safety functions. Adequate 'redundancy' is provided in systems, such that the minimum safety functions can be performed even in the event of random failures of individual components in the system. The logics and instrumentation circuits are designed to fail in the safe direction to minimise the probability of unsafe failures. Electric power requirements for safety systems equipment like pumps, valves, controls etc. are met by emergency diesel generators and batteries. Active components in systems important to safety are tested periodically.

Safety Analysis

While a very high level of safety is provided in design, the full range of possible mal-functions and failures of components and systems are analysed to predict the consequences of these failures as part of the design process to demonstrate the safety of the plant. The scope of the safety analysis covers events that are likely to occur with high frequency. The analysis of all such postulated events is documented in a Safety Analysis Report (SAR), which is subjected to thorough internal reviews and a regulatory review by the AERB, before a plant is commissioned. This provides input with respect to the provisions and procedures to handle such accidents. The accident analysis that is carried out on deterministic basis is suitably complemented by probabilistic safety analysis, which at the design stage gives insights into NPP strengths and highlights areas where changes are necessary to limit the consequences at the design stage itself.

The safety analysis has the following deliverables:

- Evaluation of the design capability to accommodate anticipated deviations from normal operation and postulated malfunctions and to establish that the operations continue to be safe.

- Establish the process requirements for design of safety systems.
- Develop technical specifications for plant operation.

In addition, a "Beyond Design Basis Events" analysis is carried out by best estimate predictions to ensure that any event beyond the design basis does not unacceptably increase the risk. This analysis provides guidance for incorporating design features which could further enhance safety and for severe accident management and emergency interventions.

Among external events, a tsunami is significant for coastal sites. Models to predict tsunami levels and resulting inundations have been developed and benchmarking of the codes is done with the participation of premier academic institutions in the country. Specifications for the stress tests of Indian NPPs are formulated with respect to assessing the cliffage effect for external events like floods, seismic activities, and station black out scenarios. Probabilistic Safety Assessment has been extended to plant shut down, external flood, and internal fire. A full-fledged testing facility to simulate any scenario for getting the requisite insights for hydrogen management has been established. This covers all the postulated incidents and accidents. Symptom based guidelines for management of accidental situation have been developed, which is a much needed operator help, in demanding situations. The safety analysis of Indian nuclear power plants is at par with international standards.

A detailed Periodic Safety Review (PSR) is carried out every ten years in accordance with the regulatory guidelines during the operating life of the plant. The safety assessments performed during PSR take into account the new developments in safety standards and operating practices, effects of plant aging, modifications, feedback of operating experience, probabilistic safety assessments and progress in science and technology. The PSR is again subjected to a regulatory review with built in multi-tier review process.

Safety in Construction and Commissioning

It is important that the NPP is constructed in accordance with the design and with the required quality of materials, processes and workmanship. This is achieved by careful execution and quality assurance in activities such as: material selection, testing, component fabrication at manufacturer works, civil construction, site erection, assembly, and commissioning. Extensive application of current non-destructive technologies is enables compliance with the prescribed requirement of the codes. Each of these activities is has written standard operating procedures and meticulous execution to ensure that the design specification is met.

After the construction is completed, a systematic commissioning process is undertaken to test and demonstrate efficiency of each system in actual performance to conform to the design. Regular plant operation is permitted by AERB only after these test results are thoroughly reviewed by the design group and AERB.

Safety in Operations

The operational phase of a nuclear plant is the final test of sound siting, design, construction and safety objectives. Well documented and approved practices, qualified & licensed operators, their training & retraining; and operational feedback are some of the measures taken for safety in operation.

The plant operation is governed by technical specifications for operations which are written down documents approved by the regulatory body. The safety limits, limiting condition of operation for different systems and their surveillance frequency are specified in the documented operating procedures. Protection system actuation set points are defined by limiting safety system settings and the set points are tested as per frequency in the technical specifications. The fall back actions and countermeasures, in case the normal configuration of, even redundant equipment, is not met, are also defined. Emergency operating procedures, covering postulated internal and external events are available and practised on simulators for training purposes. There is a system for keeping track of all unusual occurrences at operating stations. All unusual occurrences are systematically documented, reviewed and analysed to find the root-cause of the incident and for taking corrective actions by way of design modification, revision of operating procedures or replacement of a component with one with higher reliability and lower failure probability.

Qualified, trained and licensed manpower is yet another pre-requisite for safe operation. There is a programme to select and train personnel for operational and maintenance needs. The key positions in operation are manned by personnel licensed by a committee appointed by the Atomic Energy Regulatory Board. The operations are peer reviewed by other units, as well as by international agencies that ensures global co-operation for maintaining the highest standards of safety. These processes and various mechanisms of audit and reviews, mainly the regulatory audits ensure that the safety culture at the plant is kept at highest level.

Another unique feature of operations in nuclear stations is the pre-service and in-service inspection programme, during which the health of various components is checked to ensure that there are no incipient problems that could results in the failure of the component. One of the guiding principles adopted in the nuclear power programme in India is to ensure that radiation doses to plant personnel and to members of the public are kept as low as reasonably achievable (ALARA), and well within the prescribed limits. These limits, specified by the regulatory authorities, are in accordance with international standards. The measures adopted for design and operations to ensure that exposure of personnel are kept at lowest level are: a ventilation system designed to minimise airborne radioactivity in plant area; a zoning system with contamination checks of personnel and equipment at exits from radioactive/potentially radioactive areas; radiation shielding and access control of personnel to high radiation areas; and provision of lightweight and well-ventilated protective clothing and respirators for the plant personnel working in radioactive environment.

The safety policies, safety culture and the good operating practices are:

- In the operations regime, as low as reasonably achievable (ALARA) is the governing principle. While maximum limits for normal plant operation are specified in line with International commission on Radiation Protection (ICRP) recommendations, the actual exposures are kept at lowest possible levels.
- Plant operations are carried out within the limits specified in the technical specifications which are approved by the regulatory body. In addition, appropriate interlocks and administrative measures ensure adequate margins between safety limits and operating parameters. Adequate protection is provided against the operating parameters reaching the safety limits.
- Plant is operated only by qualified and licensed staff.
- Collective dose budgets are prepared by utility and approved by regulators for normal operation and special maintenance campaigns.
- Equipment and instruments are subjected to regular inspection.
- Pre service inspections generate base line data. In-service inspection is carried out according to the approved "In Service Inspections" document at all NPPs.

- NPPs are periodically subjected to a corporate safety audit, regulatory inspection and peer reviews.
- Plant operation and incidents are reviewed by Station Operation Review Committee (SORC) at the NPP level.
- Root cause analysis is carried out for all significant events.
- There are special procedures for non-standard jobs involving safety, and regulatory approval is obtained for these.

The Station Health Physics Unit maintains a close watch on the radiological status and events at plant. They also provide periodic report to the regulators. The Station Operation Safety Committees (SORC) at each of NPP regularly reviews the important safety issues. The station management keeps AERB informed of the outcome of these reviews. The Quality Assurance Group and the technical audit engineer give independent feedback to the station management on operation and maintenance of plant. The NPCIL's corporate QA group also conducts periodic audits. Each station is subjected to a peer review conducted by a group drawn from other stations of NPCIL. These are in addition to the WANO and IAEA peer reviews.

Each nuclear power plant site has an Environmental Survey Laboratory (ESL), which is independent of the nuclear power plant and reports directly to the Bhabha Atomic Research Centre (BARC) of the Department of Atomic Energy. The ESLs undertake continuous surveys of the environment in a 30 km radius around the plant by analysing the environmental matrices like: air, water; and foodstuff like cereals, milk, vegetables, fish etc. (two to three thousand samples are analysed every year) for their radioactivity content. The radiation dosage for the members of the public is computed from these measurements and is found to be only a fraction of the dose limit of 1000 μSv/year prescribed by Atomic Energy Regulatory Board (AERB) for the relevant critical group of the public. The Emergency Preparedness Plans for both on-site and off-site emergencies have been drawn up at all NPPs. These plans are subject to periodic rehearsals as part of preparedness and also include provisions for implementing counter measures in public domain.

Regulatory Safety Review

The Atomic Energy Regulatory Board (AERB) has spelt out the safety requirements in safety codes, guides and standards, issued by AERB, for sitting, design, construction, commissioning, operation and decommissioning of nuclear plants. These documents are in line with international documents and include recommendations by the International Atomic Energy Agency and other regulatory bodies. The AERB has two major review committees the Safety Review Committee for Operating Plants (SARCOP) and the Advisory Committees on Project Safety Review (ACPSR), for overseeing the safety aspects of operating plants in project/pre-operational stages. These committees carry out a multi-tier review process. The NPPs require formal clearances from AERB that authorise actions and stipulate specific conditions for all states i.e. siting, construction, commissioning, operation and decommissioning. Major steps for authorisation of a NPP by AERB are:

- Site approval
- Review and approval of design
- Authorisation for construction
- Authorisation for commissioning
- Authorisation for operation
- Authorisation for decommissioning.

The Safety Review Committee for Operating Plants (SARCOP) evaluates the performance and enforces radiological and industrial safety in all operating units. SARCOP's involvement is comprehensive, covering the safety of operating personnel, members of the public and the environment. SARCOP enforces safety standards stipulated by AERB for the operating units. It also undertakes safety surveillance in these units; reviews proposed safety related changes in design; and safety-related incidents in operating units. SARCOP appoints Unit Safety Committees for each operating station. These committees monitor, on a continuous basis, the safety status of their respective operating units and submit their reports/recommendations to SARCOP.

Global Co-operation in Nuclear Safety

International Convention on Nuclear Safety (CNS)

An international convention "Convention on Nuclear Safety (CNS)" was adopted in 1994 with inputs from different countries and national nuclear safety authorities. The purpose of the convention is to maintain a high level of safety by setting international benchmarks. The Convention is an incentive instrument. There are 75 contracting parties to this convention. The contracting parties submit national reports that are subjected to a peer review once in three years. The Convention obliges contracting parties to submit implementation reports of their obligations at meetings that are held at the IAEA. This is a national obligation.

This National Report is to be prepared in accordance with the "Guidelines Regarding National Reports under the Convention on Nuclear Safety" issued by CNS. The guidelines inter alia specify that the compliance with the national obligation under Articles 6 to 19 of the Convention must be demonstrated in the national reports to the convention. These articles cover:

- Safety measures in operating plants including assessments undertaken following any significant events
- Legislative and regulatory framework
- The national regulatory body, its mandate and duties, its structure and cooperation with other international regulatory bodies
- Responsibility of the operator
- Priority to safety
- Financial and human resources
- Quality Assurance
- Assessment and verification of safety
- Radiation protection
- Emergency preparedness
- Siting of nuclear installations
- Design and Construction of nuclear installations
- Operations of the nuclear plants

The article wise review and assessment of compliance demonstrates how the government of India fulfil its obligations under the Convention. The peer review in terms of the advance submission of the National Reports, responding to questions relating to national reports presentations are developed into report laying down best practices for each country, challenges for the future, planned actions to improve the safety and follow up from earlier review meetings. The outcome of the review process from all groups of countries is further discussed in a plenary session, for adoption by consensus, by the Contracting Parties. The national obligation to the nuclear safety is thus subjected to a comprehensive review by experts. The government of India ratified the Convention on Nuclear Safety in the year 2005. India has been participating in the review process since the CNS was ratified. This peer review has significantly contributed in enhancing the safety.

Conclusion

The overriding attention to safety encompasses the entire gamut of activities associated with nuclear power plants (NPPs): that is, siting, design, construction, commissioning, and operation. In all these activities, major efforts are devoted to ensuring safety of operating personnel, public as well as the environment. A systematic approach using well-defined principles is followed in the design of the nuclear power plants to provide the required safety features, adopting principles of defence-in-depth, diversity and redundancy. Nuclear Power Plants are constructed in accordance with the design intent, and with the required workmanship to very strict quality standards. Further the AERB enforces safety at various stages of the plant. These include site approval, and approval of design of systems and authorisations for construction, commissioning and operation and safety review during operational phase. The regulatory framework in India is indeed robust. All these measures ensure safe operation of the plants, safety of workers and members of the public and the environment.

The operational experience indicates that the radioactivity releases into the environment is well below the AERB prescribed limits. All nuclear power stations have undergone periodic AERB regulatory inspections. World Association of Nuclear Operators (WANO) peer reviews have been conducted at all nuclear power plants. In addition the IAEA peer Review (OSART) has also been introduced staring with peer review of RAPP-3&4 which was conducted in November 2012. The Environmental Management System (EMS) has been developed, implemented and certified in line with International Standards ISO-14001 in all plants. The international cooperation in peer reviews greatly to the safety enhancement through peer review and incentive mechanisms. A Management System Programme is implemented, maintained and continually upgraded for all activities and phases of the NPPs in a planned and systematic manner integrating the elements of safety, health, quality, environment, security and economics. This Management System Programme complies with the requirements laid down by the Atomic Energy Regulatory Board (AERB), the guidelines issued by the International Atomic Energy Agency (IAEA) and the good practices recommended by WANO.

Thus safety principles in Indian NPPs are accorded highest priority in all activities of siting, design, construction, commissioning, operation & maintenance and their eventual decommissioning. The documented safety standards/ codes and guidelines flow from internationally safety principles and are consistent with the best practices followed globally. The nuclear power plants are designed using defence-in-depth approach and time tested principles of redundancy and diversity. The construction is carried out in accordance with the highest safety and quality standards. India's excellent operational experience of more than 350 reactor-years has demonstrated the robustness of the safety procedures followed by Indian NPPs.

CHAPTER 5

Environmental Impacts of Nuclear Power Plants in India

G. Balachandran, K.D. Kapur and Kapil Patil

Introduction

Wide-ranging concerns have been expressed over radiation and the environmental impacts of India's nuclear power programme particularly after the Fukushima nuclear crisis in Japan. At the various newly designated NPP sites, lay people fear that their lives and the surrounding environment will be adversely affected due to construction and operation of the nuclear reactors. The public perceptions of the risks associated with NPPs have resulted in widespread opposition to construction and commissioning of the new NPPs and posed a serious challenge to the expansion of India's nuclear energy programme. The growing public debate in the country over various safety aspects of nuclear energy raises two important questions: What are the environmental risks associated with nuclear energy? And how are the risks addressed in various stages of nuclear power reactor design, construction and operations? The construction of NPPs and their safe operation requires them to pass through various stages of regulatory approvals beginning from site approval, to environmental and nuclear safety related clearances, emergency planning requirements and finally the decommissioning. The chapter seeks to analyse the risk management approaches adopted by the regulatory and operating agencies in the construction and commissioning of the NPPs in India.

The chapter has three important objectives. First, it reviews the regulatory requirements for siting NPPs. Second, it seeks to understand the evolution of legal framework for environmental impact assessment and regulations for nuclear installations in India. It seeks to assess the procedures for granting environmental clearances to nuclear installations which is crucial for minimising environmental risks from NPPs. Finally it also evaluates the process of environmental radiation monitoring surveys during the operational stages of NPPs.

Analysing the Environmental Aspects of Nuclear Energy

Globally, the expanding energy infrastructure in the form of power stations and ancillary facilities has had serious environmental consequences because of the emission of noxious gases, toxic substances and greenhouse gases. Almost 87 per cent of commercial primary energy is provided by fossil fuels.[1] Considerable progress has been made in reducing environmental pollutants emitted from coal and oil fired plants. But the environmental impact of fossil fuels remains because of increased energy use and the high cost of replacing old technologies in many countries. Most of the urban pollution in developing countries is caused by heavy reliance on fossil fuels in transportation and industries. It is estimated that a 1000 MWe coal plant produces around 320,000 tons of ash containing 400 tons of toxic heavy metals.[2] Additionally, it produces annually an average of 44,000 tons of sulphur oxides and 22,000 tons of nitrous oxides. One of the major problems of using fossil fuel is greenhouse gas emissions such as carbon dioxide (CO_2), methane and nitrous oxides (NOx) that lead to global warming and climate change. Carbon dioxide (CO_2) is the most important greenhouse gas with an annual release of nearly 25 billion tons globally.[3] The NPPs however release no harmful gases; on the contrary, the burning of fossil fuels releases more radioactivity than a nuclear reactor.

The construction and operation of mega power plants has certain impact on the terrestrial environment during the clearing of terrain for construction activities such as: loss of trees, frightened animals, noise, dust, and emissions of biocides etc. Also if cooling towers are used for the power plant, they may impact the microclimate conditions of the surrounding areas (humidity, visibility, fogging, icing and diffusion) as much as the climatic conditions may affect the operation of the cooling towers. Another important concern is the ecological impact of the release of thermal effluents which can cause changes in the habitat, in terrestrial and aquatic ecosystems. The temperature profile of a stream or a lake or mixing zones in with seawater requires continuous monitoring for analysing the impact on aquatic life.

While, these impacts are common to all mega power projects, the nuclear power plants (NPPs) mainly generate various radioactive products which can pose a serious danger to the environment and public, if released in large quantities. During the fission process, various radioactive fission products and activation components are formed in the fuel and cladding materials. Certain radionuclides which are formed in the coolant circuit of a reactor are usually released into the atmosphere as per the regulatory limits for minimising their impact on the environment. The new nuclear power reactors of latest generation (Gen II + & Gen III) are designed on the zero-discharge principle. However, a proper estimation of the radioactive releases is carried out in the pre-operation and the operational stages of the NPPs.

The most serious concern relating to NPPs, however, is the accidental release of radioactivity which can cause a serious disaster in the surrounding environment and for the population. Large scale radioactive exposures could pose serious health risks and socio-economic disruption and its mitigation require committing vast financial resources for a prolonged period of time. To avoid such disasters during the operational life-time of a nuclear reactor, it is essential that requisite safety related measures are properly implemented. Various regulatory and legal measures have been evolved over the years to minimise all the major and minor risks arising from the nuclear plants. The legal and regulatory guidelines ensure that construction and operation of NPPs do not impact adversely on the surrounding environment and population. The next section analyses in detail the various legal and regulatory guidelines for addressing environmental and radiation risks in the Indian context.

Environmental Impact Assessment Process in India

I) Siting Parameters

The technological complexities involved in the setting up and operation of NPPs require that the utmost attention be given to factors affecting the safety of public and environment. The early identification and evaluation of the environmental impact of the proposed project is necessary before any activity is undertaken. The regulation process therefore begins in the initial stages of selecting a site for setting up a nuclear plant. In India, the Atomic Energy Regulatory Board (AERB) was set up in 1983 to regulate the various stages of NPP construction and granting nuclear safety related approvals at each stage after detailed safety reviews. The AERB's consenting process for

NPPs is based on the several requirements specified in the AERB Safety Code on "Regulation of Nuclear and Radiation Facilities" and associated Safety Guides.[4] The following are the major stages of regulatory consent: siting approval, start of construction, commissioning and operation and eventual decommissioning.[5]

An important stage in the development of a nuclear power project is the selection of a suitable site on the basis of various safety parameters. Basic siting criteria are laid down to ensure that operations at the site do not cause radiological or other risks of unacceptable magnitude. The first among these is the requirement of providing an exclusion zone extending up to 1.5 km around the installation that is entirely under the control of the plant authorities. No public habitation is to be permitted in the exclusion zone.[6] Beyond the exclusion zone, is a 'sterilised zone' that extends five km from the reactor where measures are taken to ensure that there is no large increase in population, beyond natural growth.[7] Additionally, the AERB has stipulated various desirable population distribution characteristics such as (a) population density within 10 km of the site shall be less than 2/3 of the average population density in the state; (b) that there should be no population centre with more than 10000 persons within 10 km of the plant; (c) that there should be no population centre of more than 1 lakh persons within 30 km of the plant; and (d) the total population in the sterilised area should be small, preferably less than 20,000.[8]

The site review is second important stage in setting up the NPP. The selection of a suitable site is the result of a process in which impact on the environment and the risks to the population are analysed through various engineering feasibility studies and environmental impact analyses. The initial site selection survey for NPPs is carried out by the Nuclear Power Corporation of India Limited (NPCIL) based on primary information/data that satisfies the established criteria for site selection. A detailed assessment of the feasibility of incorporating site-related design inputs in the plant design is also carried out in the initial stages of selecting a site.[9] After completion of the site selection process, the NPCIL submits the pre-feasibility report to the AERB for its regulatory consent for a given site. The AERB evaluates the site mainly from the safety point of view, assesses the possibility of engineering solutions to site related problems, and ensures that the NPP could be built and operated so that radiation levels are within the limits stipulated in 'Code of Practice on Safety in Nuclear Power Plant Siting", AERB/SC/S.[10] A site is approved only when it can be adequately protected against certain external events through design based solutions.[11] The AERB has stipulated various rejection criteria, that are used to shortlist the candidate sites. Table 1 outlines the rejection criteria stipulated by the AERB and applied by the NPCIL in the site identification process. Also, the AERB guidelines require that the site be re-evaluated for future extension activities.

Table 5.1: Rejection Criteria with Respect to various Hazards for Siting of NPPs[12]

Sr. No	Hazard	Rejection criteria and SDV, if applicable
1	Earthquake	Site falling in seismic zone V as per BIS 1893
2	Earthquake	Distance from capable fault <5km
3	Earthquake	Potential for soil liquefaction
4	Earthquake/ Geological	Potential for slope instability which cannot be mitigated by engineering measures
5	Earthquake/ Geological	Potential for ground collapse/subsidence/uplift which cannot be mitigated by engineering measures
6	Geological	Possibility of formation of sand dunes
7	Aircraft Impact	Distance from small air fields <5km
8	Aircraft Impact	Distance from major air ports <8km
9	Aircraft Impact	Distance from military air fields <15km
10	Explosion	Distance from military installations storing ammunitions <10km

Note: 1: Environmentally sensitive locations like national parks, sensitive marine environment/ biota may lead to rejection of the candidate site. 2: Places with architectural/historical monuments, pilgrimage or tourists sites, within 5 kms of the site leads to rejection of candidate site.

The selection of site involves taking into consideration various factors related to topography, geology, seismology and meteorology etc.[13] Earthquakes are among the more important natural hazards to be considered in the design of nuclear installations. Various international codes like ASME (American Society of Mechanical Engineers), IEEE (Institution of Electrical and Electronic Engineers), ASCE (American Society of Civil Engineers), ANSI (American National Standards Institute) etc., are used to protect NPPs against earthquakes.[14] The seismic design requirements of nuclear power plants are more stringent compared to those for conventional structures. The Indian National Standard (Criteria for Earthquake Resistant Design of Structures, IS 1893)[15] divides the country into four seismic zones—Zones II, III, IV and V—and specifies the maximum possible earthquake in each zone (Fig. 5.1). In India, seismic guidelines and design requirements for the

Figure 5.1: Seismic Zones in India

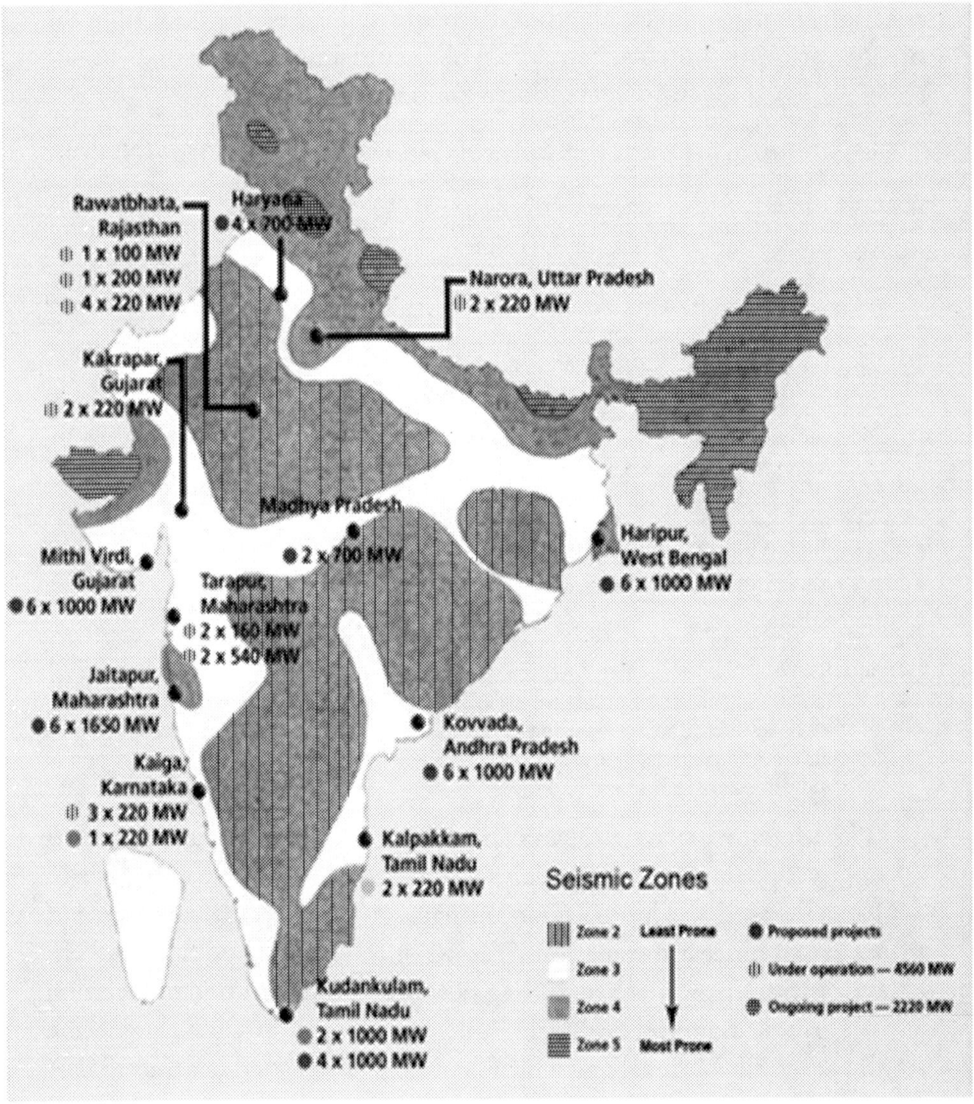

NPPs are of two types, namely: the S-1 level earthquake or Operating Basis Earthquake (OBE), and the S-2 level earthquake or Safe Shutdown Earthquake (SSE).[16] The OBE level seismic event is an event corresponding to the earthquake level that is expected to occur at least once during the life of the plant. The SSE event includes an earthquake that may be infrequent (once in 10,000 years) but is severe enough in intensity. The SEE event requires that the reactor should shutdown safely, if such an earthquake occurs. The structures, systems and components required to shut down the reactor are to be so designed as to remain functional in the event of an earthquake of a magnitude equivalent to the SSE occurs.

To mitigate the accidents caused by a major seismic event, the sites falling in high seismic zones are rejected as per AERB's rejection criteria. In the Indian context the Himalayan belt and most of Northeast region of the country that falls into Zone V is thus considered unsuitable for setting up NPPs. All the NPPs and nuclear fuel cycle facilities are located in Zones II and III with the exception of the Narora power station in Zone IV. The two units at the Kakrapar power station in Gujarat continued to operate normally, after an earthquake of magnitude of 7.9 struck Bhuj, 500 km away from the reactor in 2001. The post-earthquake inspection of the plants did not reveal any damage in any of the civil structures, pipes, pipe supports, cable trays and physical barriers provided for safety related equipment due to earthquake.[17] Several major earthquakes have therefore confirmed the earthquake resistant design and construction of NPPs in India.

II) Environmental Impact Assessment

In addition to the AERB's consenting process, the approval of a site for setting up nuclear plant requires environmental clearances as per various environmental legislations. These are granted by various central or state government agencies, as the case may be. Important legislations essential for site evaluation are: Environment Protection Act, 1986 and rules there under, Coastal Regulation Zone Notification, 1991, Water (Prevention and Control of Pollution) Act, 1974, the Air (Prevention and Control of Pollution) Act 1981, the Public Liability Insurance Act 1991 and its amendments, and Hazardous Wastes (management and handling) Rules 1989 and its amendments. An environmental clearance based on comprehensive Environmental Impact Assessment (EIA) process is a mandatory requirement for NPPs and nuclear fuel cycle facilities in India.

India has an almost 30 year experience in environmental protection for major industrial projects including NPPs. The environmental clearance for all the major projects undertaken by the government or public sector enterprises was initially introduced as an *administrative requirement* by the ministry of finance on the recommendations of Planning Commission of India. The requirement of environmental clearance based on the Environmental Impact Assessment (EIA) was introduced during 1978-79, initially for river valley projects. This requirement was subsequently extended to all the government sector industrial projects, when the ministry of finance in 1983 made it mandatory that all public sector projects requiring approval from the Public Investment Board (PIB) should obtain an environmental clearance from the department of environment (DoE). In 1985, the DoE became the ministry of environment and forests (MoEF).

In 1986, the Indian Parliament enacted the Environmental (Protection) Act. Under the 1986 Environment Act, the MoEF promulgated a notification on January 27, 1994, whereby obtaining an Environmental Clearance (EC) for expansion or modernisation of any activity or for setting up new projects listed in Schedule 1 of the notification which includes NPPs and associated Fuel Cycle facilities, became a *statutory requirement*. The January 27, 1994 notification also introduced the requirement of comprehensive EIA for obtaining an environmental clearance for the NPPs. Similarly, a mandatory 'public hearing' for setting up NPPs was introduced vide MoEF notification in 1997.

The NPP projects in India which began construction activities before the finance ministry's guidelines, stipulated in 1983, did not require an environmental clearance as an administrative or statutory requirement. These projects include: TAPS 1 & 2, RAPS 1 & 2, MAPS, NAPS and KAPS 1 & 2. The Kudankulam Nuclear Power Plant in Tamilnadu, which was conceived in 1987, obtained an environmental clearance from the MoEF on May 9, 1989, for setting up two 1000 MWe units. The nuclear power projects and nuclear fuel cycle facilities conceived after the MoEF notification of 1994 have obtained environment clearances on the basis of comprehensive EIAs and public hearings. Table 5.3 lists the status of environmental clearances for nuclear projects in India. A comprehensive EIA process ensures the early identification and characterisation of critical environmental impacts which are examined and various measures are adopted to mitigate or reduce the risks and impacts. The three main components of EIA Process are (a) EIA Study (b) Public Consultation and (c) Expert Committee Appraisal.

a) EIA Study: A comprehensive EIA study is the first step to obtaining an environmental clearance. The environmental studies are carried out in accordance with the Terms of Reference (TOR) laid down by the Expert Appraisal Committee (EAC) of the MOEF. As a primary requirement of the EIA process, the primary baseline data from the project area as well as from the area falling within the sterilised zone of the proposed project site; and secondary data has to be collected from within 30 km of the project site. The requisite data for the EIA in terms of baseline levels of radionuclides and expected levels of radioactive discharges are provide by the NPCIL. After completing the collection of data, a detailed impact analysis is carried out as mandated by the EIA Guidance Manual for Nuclear Power Plants issued by the MoEF. After completing the EIA study, the NPP project proposal is required to go through public consultation process.

Figure 5.2: EIA Clearance Schematic

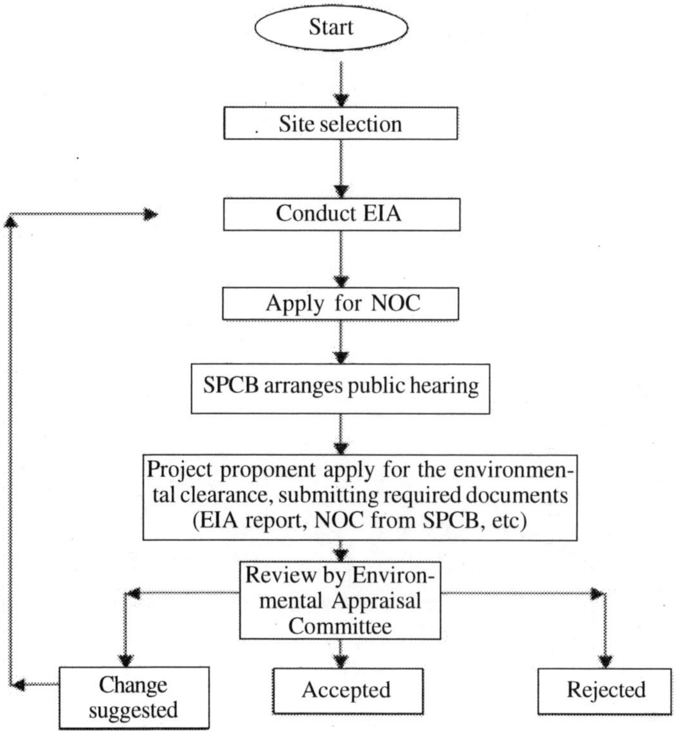

b) Public Hearing: After carrying out an EIA study, the project authorities i.e. NPCIL has to submit the study report to the State Pollution Control Board (SPCB) for arranging public consultations. "Public consultation" refers to the process by which the concerns of local affected persons and others who have a plausible stake in the environmental impact of the project or activity are ascertained with a view to taking into account all the material concerns regarding the project or activity design.[18] Public consultations in the form of a general hearing of the EIA report is mandatory for NPPs in India as per the MoEF's 2006 notification. The experience so far regarding public consultation suggests that it is an important process to involve local in the project activities and to address their concerns by providing the requisite information or clarifications, or finding additional solutions to facilitate the project. Table 5.2 shows the public hearings conducted for NPPs and Uranium mining projects so far.

Table 5.2: Public Hearings for Nuclear Projects[19]

Public hearings for nuclear projects	Date
Nuclear power plants	
PFBR, Kalpakkam	July 27, 2001
Kakrapar Units 3 and 4	April 21, 2006
Koodankulam Units 3, 4, 5, and 6	June 2, 2007
Jaitapur	April 16, 2010
Gorakhpur	July 17, 2012
Mithi Virdi	March 5, 2013
Uranium mining	
Badhurang, Jharkhand	February 25, 2004
Bagjanta, Jharkhand	September 18, 2004
Seripally,	March 3, 2005
Lambapur Pedugatta	March 3, 2005
Mouldih	August 29, 2005
Pulivendula, Kadapa	September 10, 2006
Domiasiat, Meghalaya	June 12, 2007
Thummalapalli, Andhra Pradesh	May 30, 2013

c) Environmental Clearance for NPPs: An Expert Appraisal Committee (EAC) is set up within the MOEF for granting environmental clearances to NPPs. The EAC is an important body which evaluates the environmental aspects of the project and grants environmental clearance based on the robustness of the EIA Report. The EAC is fully independent of the nuclear establishment and wields veto in the clearance process for nuclear installations.

The clearance to the nuclear project is finally granted based on the evaluation of final EIA report, and the proceedings of the public consultations including public hearing proceedings submitted by the NPCIL. Once the environmental clearance is accorded by the MoEF, their regulatory control is limited to ensuring that stipulations made in the clearance, if any, are followed by the project authorities. However, the regulatory control of AERB starts from site selection to decommissioning of the project. Other regulatory authorities like MoEF, SPCB monitor the conditions set down in the clearance letter. The controls and regulations are aimed at radiation exposure controls and ensure safety of site personnel, population living around the plant, flora and fauna.

Table 5.3: Status of Environmental Clearances for NPPs

S. No	Proposal Details	Location	Clearance Status	EC Granted	Company/ Proponent
1.	Tarapur Atomic Power Station (TAPS) I & II 2x160 MWe	State: Maharashtra District: Thane Village: Tarapur	Not Required	NA	
2.	Rajasthan Atomic Power Station (RAPS) I&II 2x220 MWe	State: Rajasthan District: Kota Village: Rawatbhata	Not Required	NA	
3.	Madras Atomic Power Station (MAPS) I&II 2x220 MWe	State: Tamilnadu District: Madras Village: Kalpakkam	Not Required	NA	
4.	Kakrapar Atomic Power Station (KAPS) I&II 2x220 MWe	State: Gujrat District: Surat Village: Kakrapar	Not Required	NA	
5.	Narora Atomic Power Station (NAPS) I&II 2x220 MWe	State: Uttar Pradesh District: Bulandshahar Village: Narora	Not Required	NA	
6.	Kaiga Atomic Power Station (KAPS) I&VI 6x235 MWe	State: Karnataka District: Karwar Village:Karnataka	Obtained	Jan 10, 1992	M/s NPCIL
7.	Tarapur Atomic Power Station (TAPS) III&IV 2x540 MWe	State: Maharashtra District: Thane Village: Tarapur	Obtained	July 26, 1990	M/s NPCIL
8.	Rajasthan Atomic Power Station (RAPS) III&IV 2x235 MWe	State: Rajasthan District: Kota Village: Rawatbhata	Obtained	Dec 4, 1986	Nuclear Power Board
9.	Rajasthan Atomic Power Station (RAPS) V&VIII 4x500 MWe	State: Rajasthan District: Kota Village: Rawatbhata	Obtained	Jan 15, 1993	M/s NPCIL

S. No	Proposal Details	Location	Clearance Status	EC Granted	Company/ Proponent
10.	Kakrapara Atomic Power Station (KAPS) III&IV 2x700 MWe	State: Gujarat District: Surat Village: Kakrapar	Obtained	Dec 19, 2006	M/s NPCIL
11.	Kudankulam Nuclear Power Project (KKNPP) I&II - 2x1000 MWe	State: Tamilnadu District: Tirunelveli Village: Kudankulam	Obtained	May 09, 1989	M/s NPCIL
12.	Kudankulam Nuclear Power Project (KKNPP) III&IV - 2x1000 MWe	State: Tamilnadu District: Tirunelveli Village: Kudankulam	Obtained	Sep 23, 2008	M/s NPCIL
13.	Kudankulam Nuclear Power Project Expansion Units V&VI - 2x1000 MW	State: Tamilnadu District: Tirunavelli Village: Kudankulam	Obtained	Dec 31 2009	M/s NPCIL
14.	Kakrapar Extension Unit 3&4 (KAPP), 2x700 MW	State: Gujrat District: Surat Village: Kakrapar		Dec 19 2006	M/s NPCIL
15.	Jaitapur Nuclear Power Project JNPP 6x1650 MW	State: Maharashtra District: Ratnagiri Village: Jaitapur		Nov 26 2010	M/s NPCIL
16.	Prototype Fast Breeder Reactor, Kalpakkam, 500 MWe	State: Tamilnadu District: Chennai Village: Kalpakkam		Apr 17 2003	M/s IGCAR

Source: MoEF's Environment Clearance Status Query Form and DAE Annual Reports.

The EIA system in India has evolved considerably as a result of numerous changes made in the original procedures to make it more encompassing and effective. The inadequacies in the Environment Clearance process have been continually addressed by making amendments to the 1994 EIA notifications and policy decisions. Keeping in view a decade of experience in the environmental clearance process and the demands from various stakeholders, the MoEF issued a revised notification on EC process in September 2006 and amended it in December 2009.

Environmental Impact Monitoring

Environmental impact monitoring is an essential element of nuclear power plant operations in India. It is carried out by the department of atomic energy (DAE) in addition to various legal requirements and also to ensure compliance with the regulatory guidelines. Various studies have shown that radiation exposure from radioactive releases from NPPs is far less than from thermal power stations. The radioactive releases, however minimal, need stringent monitoring as per the international standards. The DAE conceptualised the environmental monitoring programme for nuclear reactors, well before the commissioning of its first commercial power plant. The Environment Survey Laboratory (ESL) is established at each power plant at least two years before the commissioning of the plant. The primary aim of the environmental monitoring programme is to establish and monitor radiation levels to ensure compliance with the radiation exposure limits set for members of public by AERB. The proposal for setting up of an ESL at India's first nuclear plant, Tarapur Station was made as early as 1962. While strongly defending the proposal to set up an ESL; Dr Homi Bhabha emphatically stated that, "if there be no sanction for ESL, the reactors need not operate."[20]

In 1964, when the construction of Tarapur Atomic Power station had started, the ESL began functioning simultaneously at the Tarapur site for pre-operational surveys. In the pre-operational phase, ESLs generate baseline data on the levels of external radiation from natural sources (uranium, thorium and their daughter products, potasium-40 etc). It also measures the increase in atmospheric radioactivity from global weapon test fallouts (strontium-90, cesium-137 etc.). A detailed survey of the radionuclides in different environmental matrices (air, water, fish, silt, sediment, soil, vegetation, goat thyroid, vegetable, milk, grass, crops, fruits, meat and other dietary items) in a 30 km radial distance around the plant is undertaken. The number and type of samples and sampling frequency are optimised for each site on the basis of the type of operating facilities, utilisation of local natural resources, existence of population clusters and related demographic data.[21]

At the ESLs, three types of environmental samples are collected and analysed. The first categories of samples (drinking water, air and locally produced dietary items consumed by the public.) are directly relevant to the estimation of dosage received by members of the public. The second category of samples such as weeds, sediment, soil, grass etc is 'trend indicators' for build up of radionuclides in the environment, if any. The third category includes sensitive indicator organisms that accumulate specific radionuclides to a great extent, and these serve as very sensitive detectors or markers. For example, goat thyroid is used for detection of low levels of fresh radioactive fallout of Iodine-131. In the aftermath of Chernobyl accident, increased iodine activity in goat thyroid samples was measured at all the ESLs in the country.[22] The acceptable level for I-131 in goat's thyroid has been derived at about 500Bq/g. All ESLs carry out goat thyroid analysis in their routine monitoring programme. Thyroid samples are collected from the nearest abattoir.

Figure 5.3 shows the results of analysis of Iodine -131 in goat thyroid collected by ESL, Tarapur for the period 1967-2009. Thorough investigations have revealed no unacceptable build up of I-131 in the environment. I-131 activity was observed in goat thyroid by ESL Tarapur on few occasions which reasonably correspond to international events such as atmospheric nuclear weapon tests conducted by France and China during 1970's and 80's and also the nuclear accident at Chernobyl. The radiation dose in the vicinity of Tarapur Atomic Power Station due to operation of nuclear reactors has remained at an insignificant fraction of the AERB limit of 1mSv/year.[23]

During the pre-operational period, the ESLs also collect meteorological data, like wind speed and wind direction on a continuous basis. After the commencement of reactor operations, the ESL continues to monitor the external radiation exposure levels in the environment, to measure the meteorological parameters and to analyse the distribution and concentration of reactor related radionuclides in samples of different environmental matrices. This permits an assessment of the radioactive releases from the plant, if any. Tritium is an important radionuclide that can be released from the Pressurised Heavy Water Reactors (PHWRs). In the over-all sampling programme, therefore, considerable importance is given to the tritium measurement in the aquatic and atmospheric environment. The data collected by the environmental survey laboratories have shown that, even after thirty years of operation of nuclear power plants, the levels of radionuclides in terrestrial samples such as soil, crop, vegetation, milk, meat, egg etc., are much below the internationally accepted threshold values. The data published in various DAE publications on all the operating PHWRs in India shows only a small increase in the levels of tritium and cesium-137 in aquatic samples such as water, fish and sediments in the vicinity of discharge area show, which are too low from the health point of view.[24]

Thermo-Ecological Studies

To evaluate compliance with environmental protection laws, the DAE has carried out two important thermo-ecological studies to make a realistic assessment of the biological impact of thermal discharge on the life forms in the vicinity of the existing and/or future nuclear power plants. Co-ordinated thermal ecological studies (TES) were carried out around the currently

Figure 5.3: Iodine 131 in Goat's Thyroid

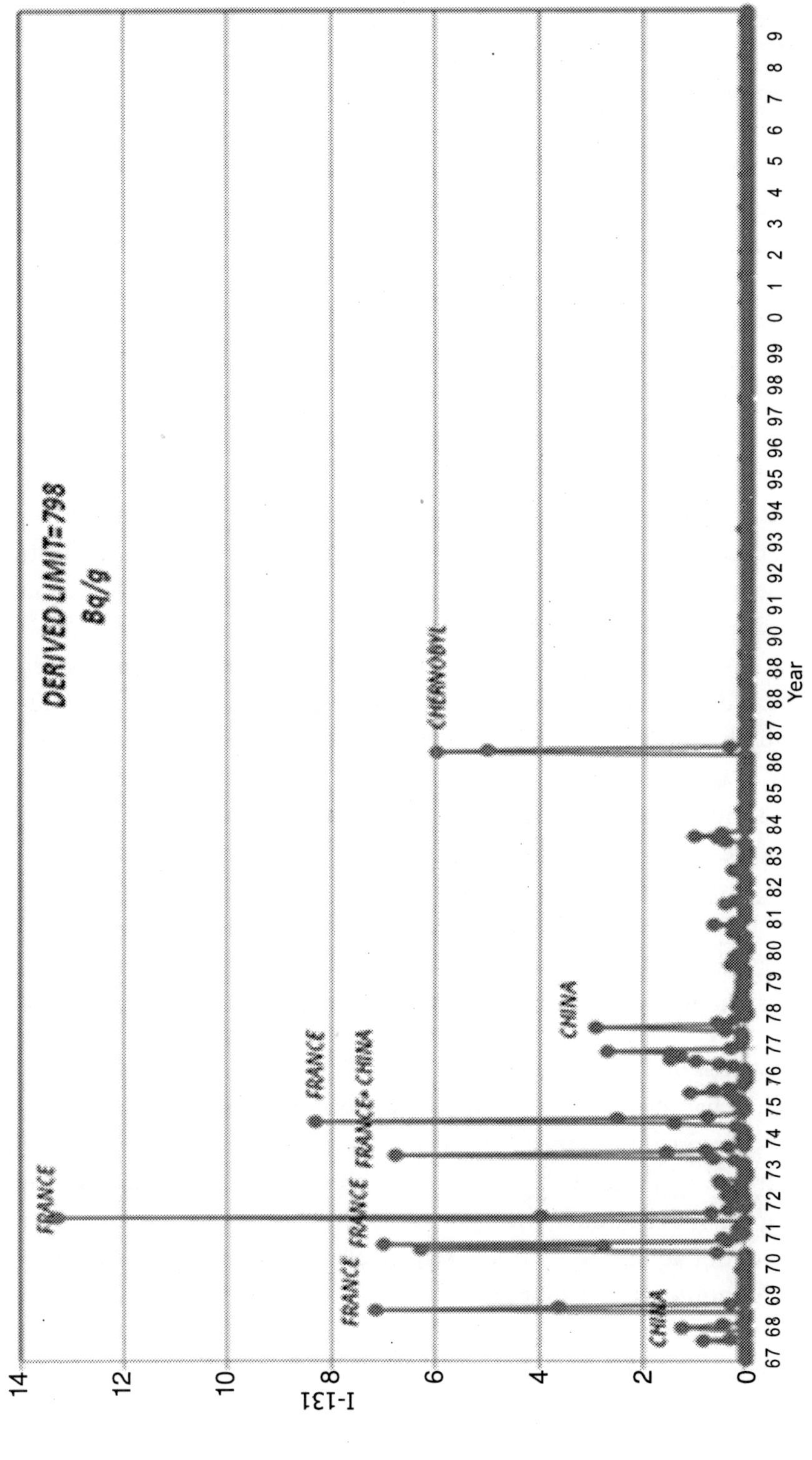

operational Kalpakkam and Kaiga plants and soon to be operational Kudankulam plant over three years. A total of 8 laboratories from different universities/institutes in India participated in these studies. At Kudankulam the study was aimed at collecting the baseline ecological information about the physico-chemical parameters of water quality, nutrient status and occurrence and distribution of major flora and fauna of the site. At Kalpakkam and Kaiga, the study assessed the effects of thermal discharge on a marine and fresh water ecosystem. The study also monitored the fish catch to assess the economic impact of thermal discharge, if any. These studies have provided a comprehensive knowledge of different aspects of thermal ecology at these sites.

The data obtained from these studies has clearly highlighted that at Kalpakkam and Kaiga sites the thermal plume distribution conforms to the MOE&F stipulations. The study noted the ability of the motile forms such as fishes and crabs to re-colonise the sites near discharge points during power plant shutdown when the thermal plume disappeared or at different times of the year when the discharge point shifted for a variety of reasons. The results of these studies uphold the concept of a "mixing zone" within which heated effluents can be allowed to mix with the water body and beyond which the life forms would be safe from possible effects of thermal discharge.[25] These studies clearly underline there are no adverse effect on marine ecology around the nuclear power plant sites. The NPCIL has also been carrying out studies of biodiversity, particularly avifauna, in and around the Indian NPPs under its Environment Stewardship Programme (ESP), in collaboration with Bombay Natural History Society (BNHS). The existing regulatory and environmental requirements in India are comparable with the international standards and are stringently applied across the programme.

Conclusion

This chapter shows that construction of NPPs in India has strictly adhered to the existing administrative as well as statutory requirements stipulated by various agencies. The existing EIA process not only helps in identifying various environmental impacts but also ensures that setting up of NPPs does not adversely affect the sensitive eco-system. The process of granting environmental clearances for NPPs in India has considerably evolved over the years and is on par with various international standards. In addition to the mandatory EIA process, the Environment Survey Laboratories (ESLs) have been carrying out extensive monitoring of radioactivity around the NPPs into various environmental matrices by way of regular monitoring surveys. The monitoring surveys by ESLs have shown no significant rise of environmental radioactivity due to operation of NPPs. The increase in radioactivity levels was registered after certain atmospheric nuclear explosions and Chernobyl accident. Additionally, thermo-ecological studies have been carried out by various regional universities which concluded, with overwhelming evidence, that the impact of thermal effluents on marine and fresh water eco-system is minimal and reversible.

NOTES & REFERENCES

1. International Atomic Energy Agency (IAEA), "Salient Points: Sustainable Development and Nuclear Power", URL: http://www.iaea.org/Publications/Booklets/Development/devfifteen.html, accessed on April 02, 2013.
2. Eric D. Graham (2008), "Environmental Impact: Nuclear Energy in Comparison with other Alternatives", *University of South Carolina,* URL: http://www.physics.sc.edu/~gothe/Study-Abroad-S08/presentations/Eric-Environmental-Impact-08.pdf, accessed on April 02, 2013.
3. Ibid, Eric D. Graham.
4. AERB Safety Code, "Regulation of Nuclear and Radiation Facilities", *No. AERB/SC/G,* URL: http://www.aerb.gov.in/AERBPortal/pages/English/t/publications/CODESGUIDES/SC-G.PDF, accessed on April 04, 2013.
5. Ibid, *Regulation of Nuclear and Radiation Facilities.*
6. AERB Safety Guide, "Criteria for Regulation of Health and Safety of NPP Personnel, the Public, and the Environment", *No AERB/SG/G-8,* June 2001, URL: http://www.aerb.gov.in/AERBPortal/pages/English/t/publications/CODESGUIDES/SG-G-08.PDF, accessed on April 04, 2013.
7. Ibid, *Criteria for Regulation of Health and Safety of NPP Personnel, the Public, and the Environment,* p. 6.
8. Ibid, *Criteria for Regulation of Health and Safety of NPP Personnel, the Public, and the Environment,* p. 7.
9. Ibid, *Regulation of Nuclear and Radiation Facilities.*
10. Roshan A.D et.al, "*Monograph on Siting of NPPs*", Atomic Energy Regulatory board, p.7, URL: http://www.aerb.gov.in/t/sj/Siting.pdf, accessed on April 06, 2013.
11. AERB Safety Guide, "Quality Assurance in Siting of Nuclear Power Plants, *No AERB/NPP/SG/S-10,* URL: http://www.aerb.gov.in/T/PUBLICATIONS/CODESGUIDES/SG-S-10.PDF, accessed on April 06, 2013.
12. Ibid, *Quality Assurance in Siting of Nuclear Power Plants.*
13. Ibid, *Quality Assurance in Siting of Nuclear Power Plants.*
14. Department of Atomic Energy, "Chapter 7 - Environmental Aspects – Radiation and Safety", in *Saga of Atomic Energy in India - 1954–2004, Commemorative Volume II,* 2003-04, p.301.
15. "Indian Standards on Earthquake Engineering", URL: http://www.bis.org.in/other/quake.htm, accessed on April 07, 2013; Bureau of Indian Standards (2002), "Criteria for Earthquake Resistant Design of Structures: Part 1, General Provisions and Buildings", IS 1893.
16. AERB Safety Guide, "Safety Guide for Seismic Studies and Design Basis Ground Motion for Nuclear Power Plant Sites", No AERB/SG/S-11.
17. AERB, "8.4.7 Indian Nuclear Power Reactors Unaffected by Earthquake", *Annual Report*

2000-2001, p. 45, URL: http://www.aerb.gov.in/AERBPortal/pages/English/t/annrpt/2000/annrpt2k0.pdf, accessed on April 07, 2013.

18. Ministry of Environment & Forests (2010), *"Environmental Impact Assessment Guidance Manual for Nuclear Power Plants, Nuclear Fuel Reprocessing Plants and Nuclear Waste Management Plants"*, Prepared by Administrative Staff College of India, Bellavista, Khairatabad, Hyderabad, March 2010, pp.2
19. Updated Public hearings for nuclear projects. See M.V. Ramana and D.B. Rao (2010), "The environmental impact assessment process for nuclear facilities: An examination of the Indian Experience", *Environmental Impact Assessment Review*, 30 (2010), p.270.
20. Ibid, *Chapter 7 - Environmental Aspects – Radiation and Safety*, p.306.
21. Bhabha Atomic Research Centre, *"Environmental Monitoring Around Nuclear Power Plants"*, URL: http://www.barc.gov.in/pubaware/enr.html, accessed on April 11, 2013.
22. K.S. Parthasarathy (2009), "Nuclear power: myths, realities", *The Hindu*, October 15, 2009, URL: http://www.hindu.com/seta/2009/10/15/stories/2009101550201400.htm, accessed on April 11, 2013.
23. Data obtained during the authors visit to the Environment Survey Laboratory at Tarapur Atomic Power Station on September 20, 2012.
24. Ibid, *Chapter 7 - Environmental Aspects – Radiation and Safety*, p.307.
25. Apte, Shree Kumar (2002), "The DAE-BRNS initiative on thermal ecological studies relevance and implications", *Bhabha Atomic Research Centre*, Mumbai, July 01, 2002.

CHAPTER 6

Nuclear Waste Management Strategy of India

P.K. Wattal

Introduction

The safe and effective management of radioactive waste has been given utmost importance since the very inception of nuclear industry in India and it covers the entire range of activities; right from handling, treatment, conditioning, transport, storage and disposal. Radioactive wastes are generated at various stages of the nuclear fuel cycle which includes the mining and milling of uranium ore, fuel fabrication, reactor operations and spent fuel reprocessing. Radioactive wastes are produced as a result of the ever-increasing use of radioisotopes in medicine, industry and agriculture. Radioactive waste management is also associated with decontamination and decommissioning activities in view of the aging of the nuclear power plants and other radio-chemical facilities including fuel fabrication and reprocessing which are in operation for more than four decades.

The objective that governs the management of radioactive wastes is protection of man and environment, now as well as in the future. In India, the necessary codes and safety guidelines for achieving this objective are provided by the Atomic Energy Regulatory Board (AERB) in conformity with the principles of radiation protection as formulated by the International Commission on Radiation Protection (ICRP). As a waste management philosophy, utmost emphasis is given to waste minimisation at all stages of design, operation and maintenance. For effective confinement, volume reduction is a key factor in the choice of technologies for use in radioactive waste management plants.

India has followed a closed fuel cycle strategy for the development of nuclear energy based on the available nuclear resources. By adopting the 'reprocessing to recycle' approach, the nuclear resources can be optimally deployed for sustained energy production for a long time. A number of reprocessing and waste management plants have been in operation in India for around four decades to treat the spent fuel from Pressurised Heavy Water Reactors. In view of this option, the concerns for the high level waste and its reduction of radio-toxicity is the key challenge. Currently, high level liquid wastes (HLLW) are vitrified. The vitrification facilities are housed and operated in indigenously designed and built hot cells with adequate remote viewing and handling arrangements.

Indian waste management facilities are co-located with waste generating facilities, i.e., the nuclear reactor, reprocessing plant and fuel fabrication facility, so as to avoid any undue radiation exposure during transportation of waste from one place to another. Extensive experience is available in the design, construction and operation of near surface disposal facilities for low and intermediate level solid waste as well as for interim storage facilities for vitrified high level waste. A programme for ultimate disposal of HLLW in deep geological repository is also being pursued.

National Framework for Radioactive Waste Management

The national policy for radioactive waste management is based upon universally a adopted philosophy of (i) delay and decay of short lived radio-nuclides, (ii) concentration and containment of radioactivity as much as practicable and (iii) dilution and dispersion of low level activity to the environment well below the nationally accepted levels which are in line with international practices.

Policy

The national policy for radioactive waste management is broadly as follows:

(a) Discharges through gaseous, liquid and terrestrial routes are as low as reasonably achievable—technical, economic and social factors are to be taken into account.

(b) Low and intermediate level solid/solidified wastes are emplaced in a near surface shallow land repository, specially engineered for this purpose.

(c) High level and alpha contaminated liquid waste from spent fuel processing and other radio metallurgical operations is virtified and stored in an interim storage facility with appropriate cooling and surveillance for a period as necessary. Thereafter, these solidified waste products are to be emplaced in a suitably engineered deep geological repository.

(d) Alpha contaminated waste not qualifying for near surface disposal is provided suitable interim storage pending its disposal in a deep geological repository.

(e) Spent radiation sources are either returned to the original supplier or handed over to a radioactive waste management agency identified by the Regulatory Body.

(f) Co-location of near surface disposal facility with the nuclear installations.

(g) In the Indian context, spent fuel is a resource material and needs to be processed for recovery and recycle of fissile material. Each reprocessing plant has, therefore, a co-located vitrification plant.

(h) The regulatory body determines the period for which active control like monitoring, surveillance, and remedial work of the repository should be maintained by the waste management agency. Thereafter, passive control like permanent markers and land use restrictions will be passed on to the central government, the agency for institutional control. Institutional control may span a period of 300 years comprising, typically, 100 years of active control and 200 years of passive control so as to allow decay of most of the radio-nuclides present in the waste rendering them innocuous.

Classification of Waste

Classification of waste is very important for safety as well as process considerations. Classification of waste into different categories is useful for their segregation, selection of appropriate treatment process, storage and disposal. As per AERB stipulation, radioactive liquid waste streams are commonly classified as exempt waste, Low Level Waste (37-3.7×10^6Bq/L), Intermediate Level Waste (3.7×10^6 - 3.7×10^{11}Bq/L) and High Level Waste (above 3.7×10^{11}Bq/L). The concentrations of long lived radionuclides and tritium are also taken into account while classifying the liquid wastes. Solid radioactive wastes are also classified as compressible or non-compressible and combustible or non-combustible depending upon the corresponding physical nature. They are further divided on the basis of type and content of radioactivity.

Basic Steps in Radioactive Waste Management

The various stages of management of radioactive waste as adopted in India are shown in Fig. 6.1.

These include waste characterisation, treatment, conditioning, storage, disposal, surveillance/monitoring, etc. The various options available for treatment, conditioning, storage and disposal of these wastes depending on their physical forms are also indicated.

Characterisation

Characterisation is applied in order to segregate radioactive materials for exemption, re-use, and disposal methods or to assure compliance of waste packages with requirements for storage and disposal.

Treatment

Treatment of waste is the initial step in waste management that occurs after waste generation. It could consist of collection, segregation, chemical adjustment and decontamination.

Conditioning

Conditioning of radioactive waste involves those operations that transform radioactive waste into a solid form suitable for handling, transportation, storage and disposal. The operation includes immobilisation of radioactive waste, placing the waste into containers and providing additional packaging. Common immobilisation methods include solidification of low and intermediate level radioactive waste in cement and vitrification of HLLW in a glass matrix.

Storage

Storage of radioactive waste involves maintaining the radioactive waste such that: (i) isolation, environmental protection and monitoring are provided, and (ii) actions involving treatment, conditioning and disposal are facilitated. Storage is undertaken for primarily technical considerations, such as storage of radioactive waste containing mainly short lived radio-nuclides for decay and subsequent release within authorised limits, or storage of high level vitrified waste for thermal considerations prior to geological disposal.

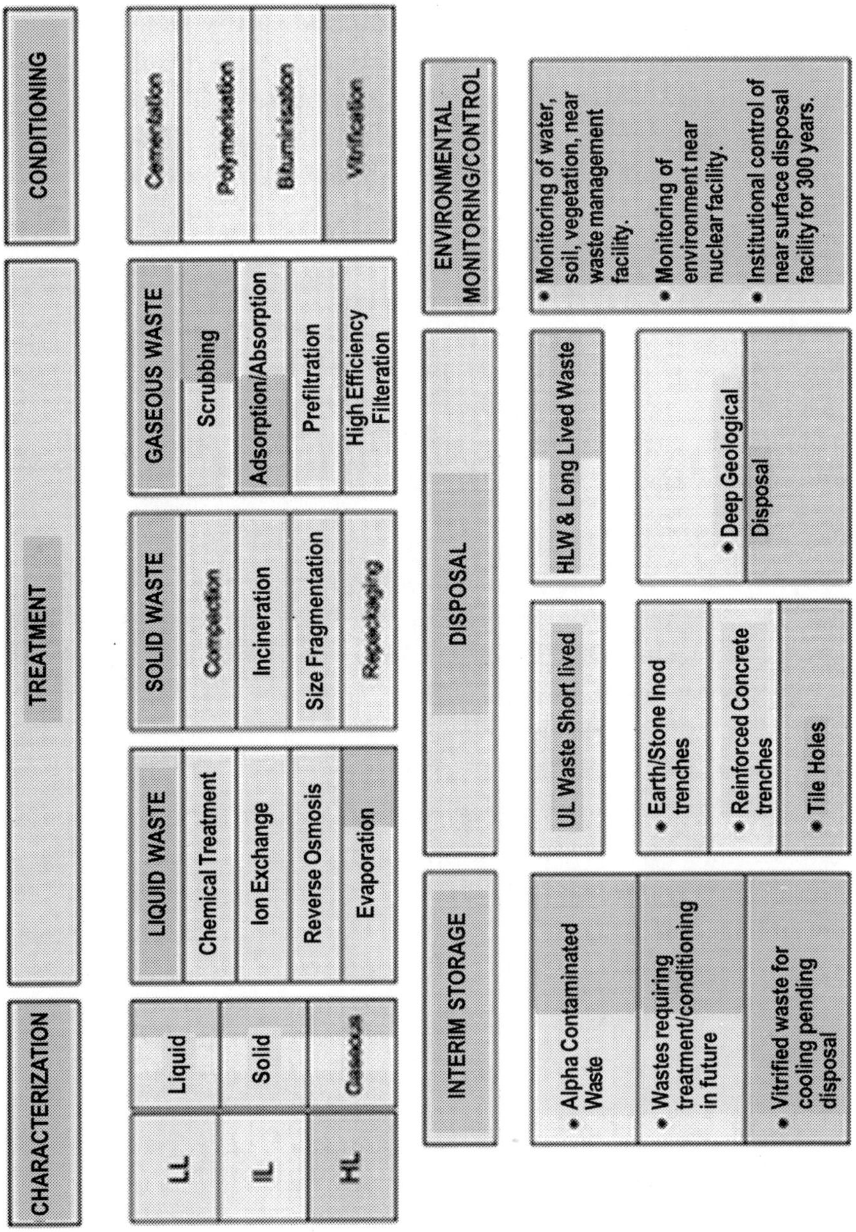

Figure 6.1: Management of Radioactive Waste

Disposal

Disposal is the final step in the radioactive waste management system. It consists of the emplacement of solidified radioactive waste in a suitable disposal facility with assurance for safety on long term surveillance and maintenance. A system of multiple barriers gives enhanced assurance of isolation and helps to minimise release of radio-nuclides into the environment. During the period when the radioactive waste is contained by the system of barriers, the radio-nuclides in the waste are undergoing decay thereby reducing the hazard over time. The barrier system is designed according to the disposal option chosen and the radioactive waste forms involved. Liquid and gaseous effluents are discharged into the environment within authorised limits. These limits are site specific and vary from coastal to inland sites. Typically, discharge limits of concentration of radioactivity in liquid effluent for coastal site are; 36 Bq/mL for gross beta and 0.36 Bq/mL for gross alpha.

Process and Technology

In view of the large variety of radioactive wastes being generated, processes used for their treatment are also diverse. Work on development of these processes and relevant technologies began in the laboratories and pilot plants in Bhabha Atomic Research Centre at Trombay in the early sixties. This work has resulted in the development of various processes and matrices for treatment and conditioning of radioactive waste.

Based on the indigenous development of process, technology, equipment and assemblies, radioactive waste management facilities have been set up at various sites, as detailed in Table 6.1 Some of these facilities have been in operation for more than 45 years. Valuable experience has been gained in the design, construction, operation and maintenance of such facilities.

Table 6.1: Some of the Waste Management Facilities in India

Site	Year of Commissioning	Nuclear facility
		Coastal
Trombay	1956	Research ReactorsFuel FabricationFuel Reprocessing Research LaboratoriesIsotope ProductionWaste Immobilization Plant (WIP)
Tarapur	1969	Power reactors (BWR, 2 × 160 MWe)(PHWR 2 × 540 MWe) Fuel Reprocessing Fuel Fabrication Waste Immobilization

Site	Year of Commissioning	Nuclear facility
Kalpakkam	1984	Research Reactor (FBTR) Power reactors (PHWR, 2 × 220 MWe) Fuel Reprocessing Research Laboratories
		Inland
Kota	1972	Power reactors (PHWR, 1 × 100 MWe 1 × 200 MWe 4 × 220 MWe) Isotope Production
Narora	1989	Power reactors (PHWR, 2 × 220 MWe)
Kakrapar	1993	Power reactors (PHWR, 2 × 220 MWe)
Kaiga	2000	Power reactors (PHWR, 4 × 220 MWe)

Treatment of Low and Intermediate Level Liquid Waste

Low level waste is generated from reactor operations, off-gas scrubbers of nuclear facilities, active floor drains, decontamination centre, laboratories, drain from change room and showers as well as secondary wastes generated during management of high and intermediate level waste. These waste streams require treatment to reduce their activity concentration to a level at which they are allowed to be discharged according to national regulations. The processes that are employed for treatment of these types of waste include filtration, chemical treatment, ion-exchange, evaporation and membrane based processes. Some of the processes presently under operation at various sites are summarised below:

Chemical Treatment

Liquid wastes with low levels ($37\text{-}3.7\times10^6$Bq/L) of activity containing ^{90}Sr and ^{137}Cs as the major radio-nuclides are treated by co-precipitation. Subsequent to precipitation, the resultant sludges are further concentrated by decantation, filtration and centrifugation. The resulting solids containing bulk of the radioactivity originally present in the liquid waste are immobilised in cement matrix prior to their disposal.

Ion exchange

A variety of sorbents and ion exchangers are used in India for the treatment of diverse types of radioactive aqueous waste streams. Conventional synthetic organic ion exchange resins are used for clean-up of spent fuel storage pool and for polishing of effluents from chemical treatment of low level waste. Amongst inorganic materials, synthetic zeolites and the clay mineral vermiculite have found industrial application. While vermiculite is used for decontamination of low level effluents, a synthetic zeolite is used for reduction of ^{137}Cs activity in spent fuel storage pool water.

A treatment process based on radionuclide separation by selective ion exchange is used for the effective management of alkaline intermediate level reprocessing waste streams. Such ion exchange plants are in regular operation at waste management plants. Indigenously developed Resorcinol Formaldehyde Polycondensate Resin (RFPR) is used in repeated loading-elution-regeneration cycles for efficient removal of ^{137}Cs which is the major radionuclide present. A chelating iminodiacetic acid resin is used for the removal of ^{90}Sr traces.

Evaporation

Evaporation is widely used for concentrating the liquid waste as it gives a very high volume reduction factor as well as a high decontamination factor. Both steam and solar evaporation methods are employed. Evaporation based on steam heating is used for waste of low volume and high activity.

Membrane Processes

Membrane based processes like reverse osmosis and ultrafiltration are also used essentially for treatment of low level liquid waste. These are generally employed in combination with other treatment methods like chemical treatment or ion-exchange process to further improve the decontamination. The volume of waste is normally reduced by a factor of ten and decontamination factor of 8-10 is achieved in this process.

Treatment of Gaseous Waste

India has developed very efficient gas cleaning techniques employing different types of wet scrubbers like ventury, dust, packed bed, cyclone separators, high-efficiency low-pressure drop demisters, chillers and high-efficiency particulate air (HEPA) filters to practically retain most of the particulate radio-nuclides.

Expertise and facilities have been developed for the design, production, testing and supply of combined particulate and iodine filters used extensively in emergency air clean-up system of nuclear power plants. In addition, ruthenium absorber employing perchloric acid treated siliceous brick (PERCOSIB), particulate respirator, charcoal impregnated sampling filter and stack sampling cartridges are also in use. Indigenously developed filter banks have been in service in almost all nuclear installations of the country. Facilities for in-situ testing of different filters at nuclear power plants have been established and are in routine use.

Management of High Level Liquid Waste

Reprocessing of spent fuel leads to generation of HLLW. Management of High Level Liquid Waste (HLLW) has to take into account the need for their isolation and surveillance for extended periods of time. To meet this long term perspective, waste isolation systems comprising multiple barriers are employed. The vitreous matrices in which these wastes are immobilised constitute the primary barrier. This along with its packaging, engineered barriers in the repository and the surrounding geology (secondary barriers) together are expected to prevent the recycling of radio-nuclides back to human environment so as to pose no hazard.

In line with internationally endorsed practices, a three stage strategy for management of HLLW has been adopted in India.

Stage 1: Immobilisation of HLLW in stable and inert solid matrices
Stage 2: Interim retrievable storage under surveillance and cooling
Stage 3: Disposal in deep geological formations

India has the unique distinction of operating vitrification plants at Tarapur and Trombay and another vitrification plant is under advanced stages of construction at Kalpakkam. An HLLW management facility is co-located near reprocessing plant so as to avoid any radiation hazard/exposure to public during transportation. The HLLW is concentrated by evaporation and stored in underground shielded stainless steel tanks as an initial processing step. These storage tanks require cooling and continuous surveillance. Transfer of this waste for the further treatment is carried out via dedicated stainless steel piping enclosed in secondary stainless steel pipes as additional containment and both are isolated from surroundings by a stainless steel box. The annulus between two pipes is continuously monitored during waste transfer operation. The stainless steel box is enclosed in a high integrity underground RCC trench which connects the two facilities.

In order to meet the challenge of the vitrification of HLLW, research and development work was started in India in the late sixties on the various areas of HLLW conditioning including formulation of matrices for immobilisation of HLLW and their characterisation. This was followed by research and development of process, equipment and assemblies to condition

HLLW into an inert and stable waste form of acceptable quality. These efforts have finally culminated in the design, construction and operation of vitrification facilities at Tarapur and Trombay.

The third vitrification facility at Kalpakkam is ready for operation. Process schematic employed for vitrification of HLW is shown in Fig. 6.2.

In India, borosilicate glass matrix has been adopted for vitrification of HLLW. Conditioned products are evaluated for various properties like product melt temperature, waste loading, homogeneity, thermal stability, radiation stability and chemical durability using advanced analytical instruments, e.g., Scanning Electron Microscope, Electron Microprobe Analyzer, X-ray Diffractometer, Inductively Coupled Plasma Spectrometer and Thermal analysis system.

Process Description

The vitrification process essentially consists of the metering of pre-concentrated waste and glass forming additives in the form of slurry/frit to specially designed vitrification melters. Our vitrification plants have operated both induction heated metallic melters as well as joule heated ceramic melters. A typical sketch of the melters deployed is shown in Fig. 6.3.

In metallic melter, main processing pot in direct contact with the molten glass is surrounded by a susceptor pot. The susceptor temperature is initially maintained at 600°C. The processing pot is made of high Ni-Cr alloy (Inconel 690) so as to withstand high temperature, oxidising and corrosive conditions. There is a freeze valve section at the bottom operable by an independent induction coil. With simultaneous concentration and calcination of waste, solid-liquid interface moves upwards.

The level of liquid waste is indicated by the temperatures sensed by the thermocouples located at different heights. The calcined mass is fused into glass at about 950°C and is soaked at 950-1000°C for eight hours to achieve homogenisation. The molten mass is then drained into stainless steel AISI 304L canisters by operating the freeze valve. The canister filled with vitrified waste product (VWP) is allowed to cool slowly in an insulated assembly. This is then welded remotely by the Pulse Tungsten Inert Gas method. A photograph of welding of lid of the canister containing vitrified waste product by remote welding machine is given in Fig. 6.4. In the ceramic melter, melting of the glass is achieved by joule heating wherein current between the

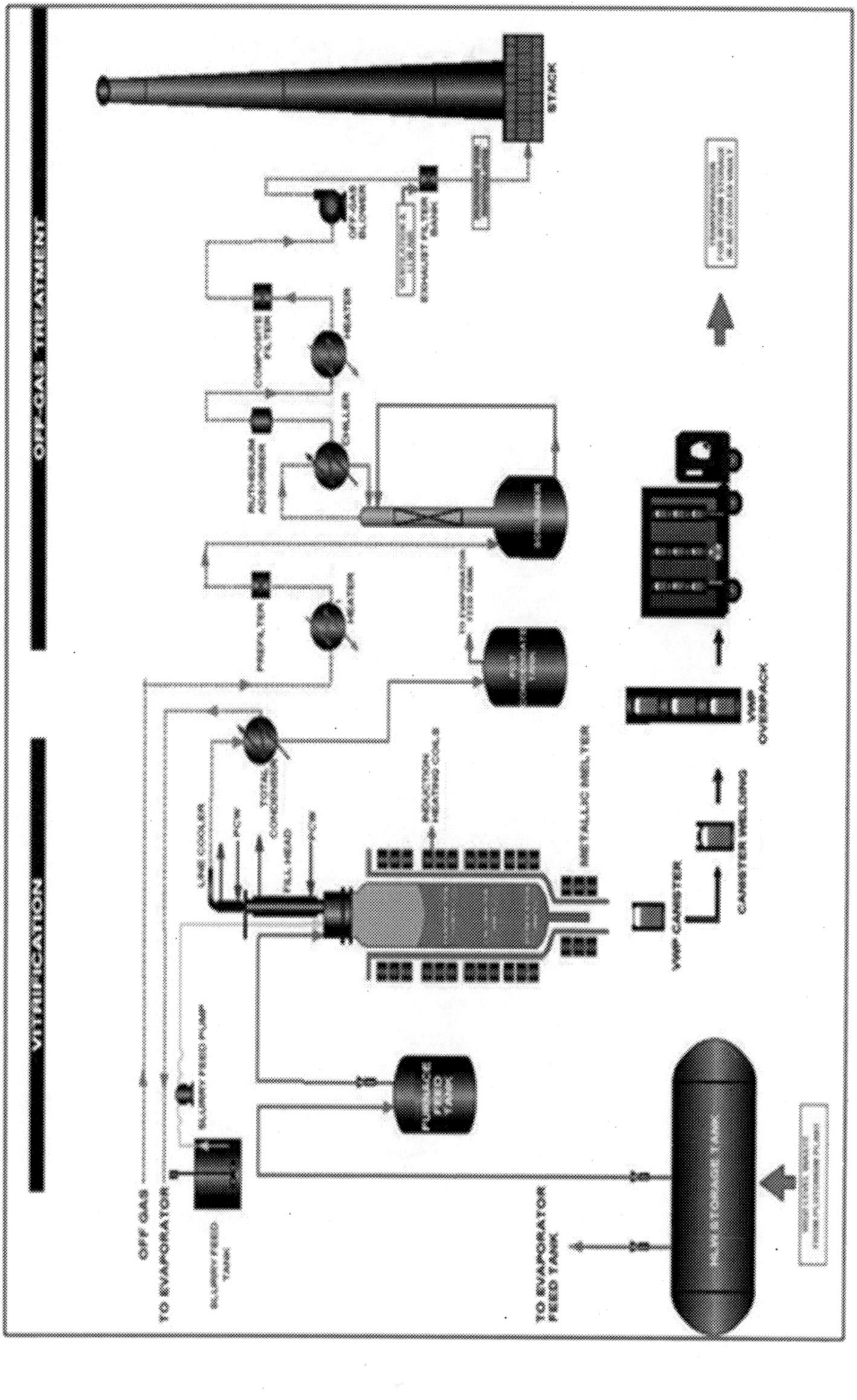

Figure 6.2: Vitrification Process for High Level Liquid Waste

Figure 6.3: Induction Heated metallic Melter

Figure 6.4: Remote Welding Machine

two electrodes within the refractory melter cavity is utilized for the formation of molten vitreous mass.

The multi-cell multi-compartment concept has been adopted at the Waste Immobilisation Plant, Trombay, so as to facilitate segregation of equipment and ease of maintenance. The process cells are equipped with state-of-the-art remote handling systems. Some of the main remotisation gadgets are servo-manipulators, close circuit television cameras and remote welding machine. A photograph of servo-manipulator station from where all remotisation gadgets are operated is presented in Fig. 6.5.

Figure 6.5: Servo-Manipulator Workstation

An elaborate off-gas cleaning system consisting of condenser, scrubber, chiller, demister and absolute HEPA filter is used to treat the gas before discharge through a 100 m tall stack to the atmosphere. Vitrified waste canisters are further enclosed in secondary stainless steel containers called over packs. The over packs may contain radioactivity up to 10^6Ci generating about 3-4 kW of decay heat and they need to be cooled continuously to maintain centre line glass temperature below softening temperature to minimize devitrification.

Management of Organic Liquid Waste

Organic liquid waste is generated from reprocessing plants in the form of spent PUREX solvent (30% tri-butyl-phosphate in diluent) which has undergone chemical and radiolytic degradation with repeated use. A plant

(Fig. 6.6) based on indigenous technology is used to treat the spent solvent by the 'alkaline hydrolysis' process. The treatment enables the recovery of diluent virtually free of activity and Tri-Butyl-Phosphate. The aqueous bottom arising from alkaline hydrolysis process is immobilised in cement. An incineration facility (Fig. 6.7) is also available for thermal destruction of non-recyclable diluents.

Figure 6.6: Alkaline Hydrolysis Plant

Figure 6.7: Organic Liquid Waste Incinerator

Management of Solid Waste

Radioactive solid wastes produced at different stages of nuclear fuel cycle cover a wider range of materials, sizes, shapes and degrees of contamination.

These wastes are categorised depending on the radiation field, concentration and type of radioactivity. They are segregated as compressible or non-compressible and combustible or non-combustible. A major portion of the total solid waste has low activity and is either combustible or compressible. Specially designed incinerators are employed for burning the combustible wastes achieving a volume reduction of about 50. A hydraulically operated baling press is used to compress low active non-combustible waste to obtain a volume reduction of five. An incinerator operational at Kalpakkam is shown in Fig. 6.8.

Figure 6.8: Incineration of Solid Radioactive Waste at Kalpakkam

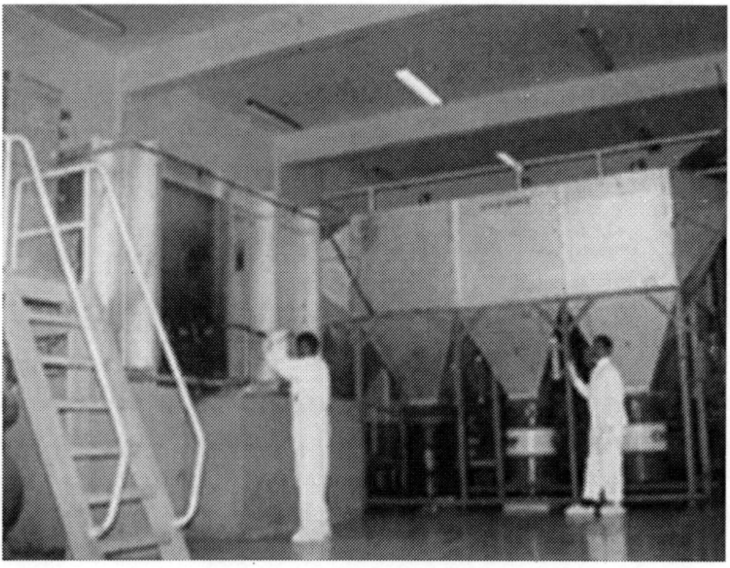

Management of Spent Radiation Sources

Radiation sources of various types and strengths, containing ^{60}Co, ^{90}Sr, ^{99}Tc, ^{99}Mo, ^{125}I, ^{131}I, ^{137}Cs, ^{192}Ir and ^{226}Ra, are used in hospitals, industries and research institutes all over the country. The strength of spent sources varies from millicuries to thousands of curies depending upon the area of application. After their utility period is over, the spent sources are immobilised in cement grout to make them suitable for disposal in engineered near surface disposal facilities. Waste management facilities located at Trombay and Kalpakkam are the nodal centres for storage/disposal of spent radiation sources.

Remote Handling Technology

Remote handling, robotics and automation are essential for operation and routine maintenance of waste management plants handling high radioactivity. The DAE with the participation of Indian industry has developed all the equipments essential for such applications and is self-sufficient, at present in this regard. Chemical processing operations in waste management plants are carried out in "hot cells" with heavy shielding (approximately 1.5 m of concrete) around the equipment. The general philosophy of remote handling is to employ a "work station concept" consisting of a radiation shielding window (RSW), master slave manipulator (MSM) and a cell crane. All the equipment required for these duties is manufactured and available on a regular basis in India and have been standardised for such applications. Models of MSMs of capacities ranging from 9 to 45 kg have been indigenised with electrical indexing capabilities to improve volume coverage. Master Slave Manipulators are available in single and modular models with partial and full internal shielding. Articulated Master Slave Manipulators required for lead/steel hot cells have been developed and are available in 5 kg capacity. Indigenous capability has also been developed and tested for a 15 kg capacity bi-lateral force reflecting electrical servo manipulator, for employment in high active cells with corrosive environment. Customised cell cranes are available indigenously in stainless steel material of construction with remote maintenance features and provisions for emergency operations. Technology for power manipulators has also been developed and is in use. Bridge mounted power manipulator of 100 kg capacity with 7-degrees of freedom namely X, Y, Z, azimuthal, roll, pitch & gripping configuration and telescopic boom has been designed and manufactured for reprocessing and waste vitrification plants.

Viewing in hot cells is accomplished either with the help of direct viewing systems such as radiation shielding windows (RSW) or radiation resistant CCTV systems. On both these fronts indigenous capability has been developed. Technology for design/manufacture of oil filled and dry type RSWs is established. Indigenous radiation tested CCTV systems are being employed for in-service inspection of reactor components.

Decommissioning of Nuclear Installations

The AERB has made it mandatory for all nuclear installations to make provisions for in-situ decontamination and de-commissioning provisions from design stage till the end of operational phase. India has considerable experience in this field and has successfully decommissioned/refurbished research reactors, major systems of power reactors and also radio chemical installations including reprocessing plant. A major activity as a part of refurbishment of operating power reactors at Rajasthan and Madras covers the replacement of coolant channels. The replaced channels, end fittings and other components are volume reduced and disposed of after conditioning. A view of the coolant channel cutting machine is shown in Fig. 6.9.

Figure 6.9: A View of Coolant Channel Cutting Machine

Development activities are underway for cutting and size reduction of metallic components using techniques like under-water plasma and lasers as part of the programme for conditioning of decommissioning waste. Technologies for melting and super compaction are being developed to achieve further volume reduction. India has also a unique experience of having dismantled and decommissioned joule heated ceramic melters from vitrification plants.

Interim Storage of Vitrified Waste

High level vitrified wastes are characterised by decay heat and need to be cooled to a level where transportation and disposal in a geological repository becomes viable and economical. This period of cooling is also used to generate data on product behaviour, under constant surveillance and monitoring. This data is essential for prediction of long term behaviour of the vitrified products. These requirements necessitate interim storage of over packs spanning over 30 years and more. In India, one such storage and surveillance facility co-located with a vitrification plant is already operational at Tarapur. This facility has a capacity for storing nearly 1700 over packs with an inventory of nearly 80,000,000 TBq of radioactivity. The facility consists of an underground outer vault (hydraulic) of dimensions 74.5m × 31.5m × 5.2m and houses two inner vaults (thermal) of sizes 34.0m × 25.2m × 2.5m each. The over packs are suspended vertically from the top slab. A concrete roof of 1.2 m thickness provides adequate radiation shielding. The wall temperature in the thermal vault is expected to be 90-110°C when the vault is full.

Removal of decay heat (design value of 3.8 kW/over pack) from the over pack is achieved by natural convective ventilation induced by a 100 m high stack. The air cooling system has been designed on the basis of storage unit geometry, array design, filling pattern and stack dimension. This is an inherently self-regulating system and takes care of the changes in decay heat. The cooling system ensures that the temperature within the vitrified waste product, under no circumstances, exceeds softening point of the vitrified mass. A schematic of the air-cooled Solid Storage and Surveillance Facility, Tarapur is shown in Fig. 6.10.

Figure 6.10: Schematic View of Air-Cooled Solid Storage Surveillance Facility, Tarapur

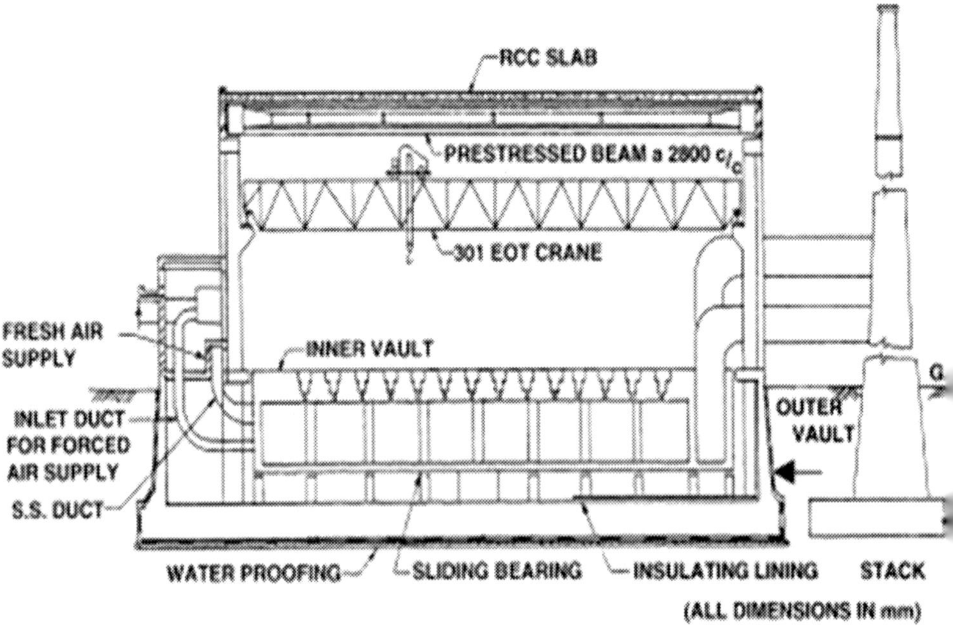

Research and Development in Waste Processing Technology

India has decades of experience and expertise in the safe management of radioactive waste. It is recognised that the technologies, currently in use are adequate, but sufficient scope exists for improving these technologies so as to enhance process performance and meet the future challenges. Development and induction of cross-cutting technology has to be undertaken not only to meet the challenges, posed by approach to near-zero discharges, but also to ensure the recycling and recovery of valuable resource from these wastes to create a positive impact on the environment. Besides, Waste Management in India has to meet the requirements of Advanced Fuel Cycles for Fast Breeder Reactors and Advanced Heavy Water Reactors.

Low and Intermediate Level Waste

High volumes and low activity levels characterise low and intermediate level wastes. Hence, attention is focused on technology development that can lead to downsizing of equipment, effective decontamination and minimisation of secondary wastes. Technologies playing a vital role in this aspect, include synthesis and use of specific sorbents, ultra filtration, and management of spent solvents by alkaline hydrolysis, development of advanced oxidative techniques for destruction of organic ion exchange resins, etc.

As already indicated, the effective treatment of alkaline intermediate level reprocessing waste is now possible because of the development of cesium-selective Resorcinol Formaldehyde Polycondensate Resin (RFPR). The resin is manufactured in-house. Adoption of this treatment also enables the recovery of ^{137}Cs in kilocurie quantities. In a plant being set up at the Waste Immobilisation Plant, Trombay, the recovered ^{137}Cs will be further processed and immobilised in a vitreous matrix for use as the radiation source in blood irradiators.

The removal of nitrates present in reprocessing effluents is essential before such effluents can be discharged into the environment after treatment for removal of radioactivity. Laboratory and bench scale studies have shown that biological denitrification under anoxic conditions in the presence of an added carbon source is an effective process for the destruction of nitrates. A pilot facility treating about 2M^3/day of nitrate bearing effluents has been demonstrated based on this bio-process. Nitrates from 3000 ppm have been brought down to 20 ppm. At the same time chemical denitration is also being attempted for high nitrate bearing alkaline wastes

High Level Waste

Globally emerging vitrification technology based on the cold crucible induction melting is being studied to address various requirements such as high temperature availability, high waste loading, high specific capacity, compatibility with new matrices such as glass-ceramics, etc. A full scale cold crucible melter has been designed, developed to establish melter start-up procedure, simulated liquid waste feeding and melter operational stability. Very few countries have access to this technology. Studies on the long-term evaluation of vitrified waste product under simulated conditions have been conducted in especially designed hot cells at Solid Storage and Surveillance Facility, Tarapur. Effect of components of repository like granite and back-fill material as well as corrosion products on leach rate of vitrified product are being studied. Some of the leach rate experiments continue for years together. The presence of secondary phases on the waste product and their effect on chemical durability is also being studied.

Partitioning of High Level Waste

Waste management being a case of sustainable development, there is a constant endeavour for waste volume reduction. Use of partitioning technology not only addresses reduction in radio toxicity but also waste volume reduction. Strategy for the management of high level liquid waste is to partition minor actinide and recover and reuse useful fission products.

Separation of Heat Generating FPs

Separation of heat generating FPs like Cesium137, and strontium 90 from high level waste leads to high loading of waste oxides on to glass in view of the reduction in the center line temperature of the vitrified waste canisters not exceeding glass transition temperatures. This will also ultimately lead to a drastic reduction of the thermal loads of the repository, translating into better space utilisation. A separate Near Surface Repository shall have to be provided for emplacement of these non-alpha high heat generated wastes.

Partitioning of Minor Actinides

Partitioning of minor actinide will not only result in the reduction of radio toxicity in the deep geological repository but will also be reflected in reducing the depths for the geological repository and public confidence. The partitioned minor actinides and some of the long lived fission products can be subjected to transmutation in dedicated Accelerator Driven systems (ADS) or recycled in Fast Breeder Reactors (FBRs). The Partitioning programme to be implemented should be as per the reference flowchart given in Fig. 6.11 and

Figure 6.11: Partitioning Reference Diagram

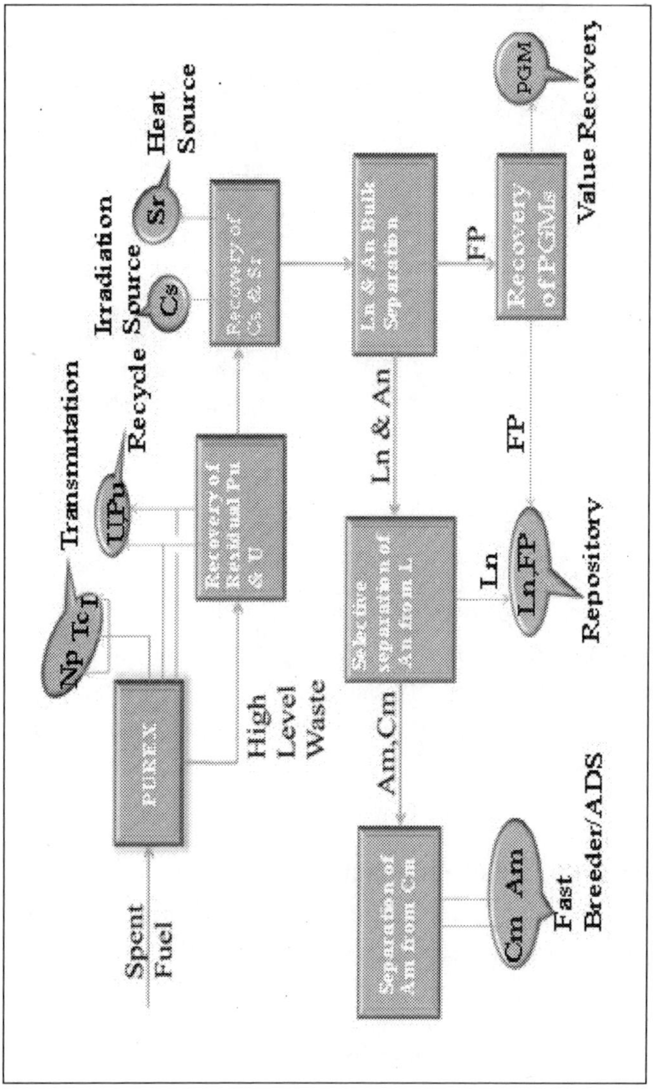

Figure 6.12: Partitioning Facility for High Level Waste under Commissioning at Tarapur

are to be included in all the future integrated plants. The demonstration facility for partitioning of high level waste set up at Tarapur is given in Fig. 6.12.

Development and Deployment of Novel Solvents

Implementation of partitioning is a direct consequence of the industrial scale availability of novel solvents. The solvents have to address good selectivity, extractability and reversibility and their waste management after they are spent. This scenario will result in the reduction of vitrified waste canisters which have a direct impact on the size of deep geological repository.

Storage and Disposal of Radioactive Waste

The present practices in India for the storage and disposal of various categories of solid waste are briefly described below. Methodologies adopted for safety and performance assessment are also discussed.

Near Surface Disposal Facilities (NSDF)

As a national policy, each nuclear facility in India has its own Near Surface Disposal Facility co-located. There are seven NSDFs currently operational within the country. These are located at Trombay, Tarapur, Kalpakkam, Kota, Narora, Kakrapar and Kaiga. These NSDFs have to address widely varied geological and climatological conditions. The various disposal modules currently adopted in NSDFs are shown in Fig. 6.13.

(i) Stone Lined Earth Trenches (SLT)
(ii) Reinforced Concrete Trenches (RCT)
(iii) Tile Holes (TH)

Figure 6.13: Battery of RCC Trenches

Figure 6.14: Battery of Retrievable Tile Holes

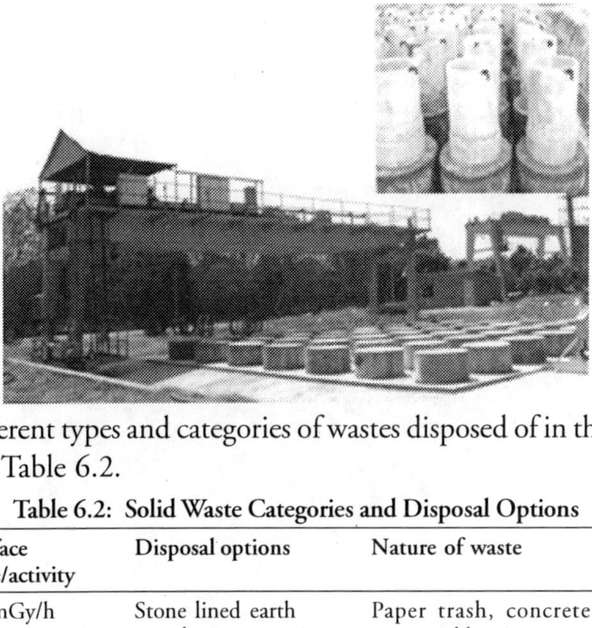

The different types and categories of wastes disposed of in these modules are listed in Table 6.2.

Table 6.2: Solid Waste Categories and Disposal Options

Category	Surface dose/activity	Disposal options	Nature of waste
I	<2 mGy/h	Stone lined earth trenches	Paper trash, concrete chips, cotton mops, rubber items etc.
II	2–20 mGy/h	RC trenches	Contaminated equipment, hardware and filters
III	20–500 mGy/h	RC trenches	Conditioned/processed concentrates, sludges, spent resins
	>500 mGy/h	Tile holes	Hardware from reactors, highly contaminated equipment, conditioned spent resins, spent radiation sources etc.
IV	Waste bearing alpha activity (<4000 Bq/g)	RC trench and Tile holes	Solidified alpha waste with β, γ activity

These modules are generally below ground. However, depending upon the local geo-hydrological conditions, these could be partly or completely above ground. Over the years, considerable experience and expertise has gone in refining and improving the design and construction of these disposal modules.

Two-tier Disposal System

In view of the growing waste generation due to expansion of nuclear programme in India, studies have been performed on two-tier disposal system and above ground RCC dyke for low level radioactive solid waste. A typical two tier disposal module has been designed to optimize the land utilisation in NSDF.

In-situ Radionuclide Migration Studies at NSDF

In-situ radionuclide migration experiments (Lysimeter experiment) are conducted to validate the radionuclide migration mathematical modeling results and to study the performance of natural barrier in case of any release from the disposal modules. Radioisotopes ^{3}H, ^{137}Cs and ^{60}Co are used as a contaminant to validate the mathematical model results. A photograph of experimental setup along with instrumentation is shown in Fig. 6.15.

Figure 6.15: Lysimeter Along With Instruments Installed at NSDF

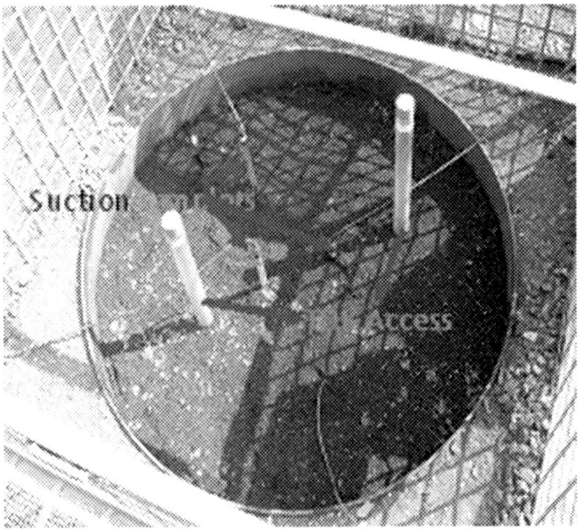

Surveillance

Provisions for monitoring and surveillance are incorporated in the design of the disposal facility. Boreholes - 4 to 7m - deep are provided at appropriate locations and the groundwater samples are monitored periodically. Soil and vegetation samples from the site are also periodically investigated for any uptake of radioactivity. Radiation survey of the entire site is carried out at predetermined intervals. The RC trenches are provided with inspection pipes to monitor the inside condition after closure of the trench. The entire disposal site is totally closed by a physical security wall to avoid unauthorised access.

Performance Assessment

The performance assessment of NSDFs is systematically undertaken through field investigations and predictive modeling. This assessment is based on a good physical understanding of the waste forms, waste packages, backfill materials and the engineered barriers. Models have been developed to predict the probability of failure as a function of target lives for various safety indices such as concrete cover thicknesses, climatic factors, maintenance period for the structure, water to cement ratio, water proofing etc. The modelling studies for a typical RC trench under limiting conditions have a predicted minimum service life of nearly 240 years. This period is sufficient for the decay of major radionuclides like ^{90}Sr, ^{137}Cs to 99 per cent of their original contribution. Thus, the source term itself reduces substantially.

Safety Assessment

Safety assessments have been carried out on each of the NSDFs. These take into account the nature of the facility, the radioactive inventory, geological, the hydro-geological and geochemical behaviour of the site, pathways and possible scenarios for release and transport of radioactivity. Mathematical models were generated on the basis of these inputs. Release and transport calculations through various barriers were determined to arrive at radiological dose estimates.

Estimates of the radioactive inventory for a typical costal NSDF catering to an operating power reactor, a reprocessing plant and a fuel fabrication facility are of the order of 30,000 TBq. The major radionuclides of concern are ^{137}Cs, ^{90}Sr and ^{60}Co. Based on this inventory, the maximum possible radiological dose at a distance of 500 m is almost negligible after a period of 730 years from the closure of the disposal facility. This dose is contributed mainly by ^{90}Sr through groundwater drinking pathway. Maximum possible radiological dose by human intrusion scenario is of the order of 10^{-6}Sv/ annum by the excavation-inhalation pathway. Since interim storage is for a period of 30-40 years, the need for the repository to be operational is beyond 2040.

Summary

Waste management facilities at various nuclear installation sites are operating safely and successfully for more than four decades. The Indian experience of management of nuclear waste from power plants, fuel reprocessing and allied installations is rich and comparable with international practices. By suitable treatment and conditioning of waste it has been demonstrated that prime objective of safety of environment is fully achieved. Elaborate surveillance ensures safe performances of the systems.

Among the many possible forms, borosilicate glass has been adopted for immobilisation of HLW. India has developed sufficient expertise in this advanced technology with respect to matrix development, process design, construction, commissioning, operation and remotised maintenance. As far as the technology for waste conditioning is concerned, metallic melter technology and ceramic melter technology have been adopted successfully. New generation technology based on cold crucible melting is under development and demonstration. A valuable base of human resources consisting of scientific and technical personnel well versed in design, construction, operation and maintenance aspects of the vitrification facilities and related R&D systems has been built up. Expertise has been provided to national and international institutions.

India has extensive experience in the handling, storage and disposal of low and intermediate level solid waste in near surface disposal facilities (NSDFs). Besides, there is adequate expertise in site selection, characterisation and safety analysis for NSDFs. Technology has been developed for the design, construction and operation of interim storage facility for vitrified waste products under surveillance. A large-scale site selection programme has been initiated to identify a few suitable geological domains for locating a repository for high and long lived radioactive wastes. Presently granites have been studied extensively as a natural barrier. Long term evaluation of vitrified high level waste under geological conditions is being pursued in India. This will go a long way in demonstrating the safe containment of waste in deep geological repositories, with no impact on the environment.

CHAPTER 7

Radiation Releases from Nuclear Power Plants in India

G. Balachandran, K.D. Kapur and Kapil Patil

Backdrop

Nuclear energy generates fewer green house gases compared to other fossil energy sources such as thermal energy. Nuclear power stations are designed to deliver a reliable level of power over long periods of time and at competitive tariffs. For this reason, nuclear power has been embraced by many countries worldwide. However, the rise in the use of radiation and radioactivity for industrial, medical and power generation purposes is widely perceived to have enhanced the risk of radiation exposure for human beings. The radioactive releases from nuclear tests and reactor accidents have become a matter of serious concern for scientists, statesman and lay people alike. India's first prime minister Pandit Nehru as early as in 1952 commissioned a major study to analyse the impact of the Hiroshima and Nagasaki atomic bombing on human health.[1] In 1954, Prime Minister Nehru in the United Nations General Assembly called for an immediate end to all nuclear explosions expressing serious concerns over the effect of radioactive fallout on human health.[2]

In order to understand the biological effects of radiation; the General Assembly on December 3, 1955 unanimously approved the US sponsored resolution 913(X), which established the United Nations Scientific Committee on the Effects of Atomic Radiation (UNSCEAR) to collect and evaluate information on the levels and effects of ionising radiation.[3] UNSCEAR has published several reports on scientific observations and experiments relevant to the effects of ionizing radiation on environment. Other than UNSCEAR, several scientific bodies have carried out elaborate studies to understand the specific biological effects of ionising radiation. This includes, the International Commission on Radiological Protection (ICRP)[4], the Committee on the Biological Effects of Ionising Radiations (BEIR) of the US National Academy of Sciences[5] and the National Radiological Protection Board (NRPB), UK. Because of these concerted efforts, biological effects nuclear radiation have been more thoroughly studied than any other potentially harmful occupational environmental agent.

While, the gamut of man-made uses and applications of atomic radiation is vast; this chapter specifically focuses on radiation releases from nuclear power plants and fuel cycle facilities in India. The safe operation of nuclear fuel cycle installations is essential for the long term safety of the people and the environment in surrounding areas. A primary concern associated with nuclear plants is the accidental as well as routine releases of radioactive materials into the atmosphere, which could pose long term risk to environment and public health. While, the probability of such large scale reactor accidents leading to widespread radioactive contamination is minimal, the very possibility of such accidents occurring at the NPPs has been a major source of fear in the public. The large-scale radiation releases from Chernobyl and Fukushima accidents have created strong fears about nuclear energy. There are also widespread misgivings about the routine releases of radioactive materials during the operation of nuclear power plants and their possible impact on public health.

A proper perspective of radiation releases from nuclear power plants can help assuage some of these concerns. This chapter aims to analyse the various effects of low level and high level radiation on public health as established by various studies conducted by the ICRP and the UNSCEAR. Second, it evaluates the health impacts of major and minor accidents at various nuclear and radiological facilities all over the world. Finally, it evaluates the radioactive releases from Indian NPPs and provides an assessment of the health impacts as documented by the extensive epidemiological studies on radiation and non-radiation workers and their families.

Radiation—A Scientific Perspective

Radiation is widely present in the earth's atmosphere and is classified mainly as ionising and non-ionising radiation. Of the two, ionising radiation is a central concern for scientists and health physicists because of its impact on public health and environment. Ionising radiation is popularly termed a '*double-edged sword*' since it causes cancer and at the same time it is used for treating the cancer. Ionising radiation contains sufficient energy to penetrate any given material and causes ionisation within that material leading to changes in chemical structures. When the human body is exposed to large doses of ionising radiation it can damage tissues in human body by damaging critical chemicals within the cells such as the DNA. Any irreversible chemical and biological damage in the cell can be harmful for human health.[6] The principal ionising radiations are: alpha, beta, gamma, x-rays, cosmic rays, and neutrons.[7] These ionising radiations are spontaneously emitted because of the decay of various naturally present radionuclides in the earth's crust as well as in the atmosphere.

Radiation Measurement

Radiation emitted by a radioactive substance is absorbed by any material it encounters, dead material or living cells. Every kilogram (kg) of material absorbs some energy (joule or J). That unit, the J/kg, is used for the measurement of the absorbed dose. In radiation protection, that unit is called the gray (Gy). The degree of biological risk caused by different types of radiation can be calculated by multiplying the absorbed radiation dose (Gy) by a radiation weighting factor. The lowest is 1 for gamma radiation and the highest 20 for alpha radiation. When an absorbed dose is multiplied by the appropriate radiation weighing factor, the resulting quantity is the equivalent dose, measured in Sieverts (Sv). The radiation dose is expressed in Sieverts (Sv). It represents the amount of radiation energy absorbed by living tissue and the extent of biological effects involved. All doses given in Sv or mSv are comparable regardless of the type of radiation involved. As the Sievert is a fairly large unit of measurement, the millisievert (mSv) is frequently used.

Alpha radiation is a heavy, very short-range particle that is unable to penetrate human skin. Alpha-emitting materials can be harmful to humans if the materials are inhaled, swallowed, or absorbed through open wounds. Examples of some alpha emitters: radium, radon, uranium, thorium.[8] *Beta radiation* is a light, short-range particle and may travel several feet in air and is moderately penetrating. Beta radiation can penetrate human skin upto the 'germinal layer', where new skin cells are produced. Beta-emitting contaminants may be harmful if deposited internally. Some beta emitters, however, produce very low-energy, poorly penetrating radiation that may be difficult or impossible to detect. Some of the difficult-to-detect beta emitters are hydrogen-3 (tritium), carbon-14, and sulfur-35. The pure beta emitters are strontium-90, carbon-14, tritium, and sulfur-35.[9] *Gamma radiation and X-rays* are highly penetrating electromagnetic radiation. Gamma radiation or x rays are able to travel many feet in air and many inches into human tissue. They readily penetrate most materials and are sometimes called 'penetrating' radiation. X-rays are like gamma rays and they too, are penetrating radiation. Sealed radioactive sources and machines that emit gamma radiation and X-rays respectively constitute an external hazard to humans. Examples of some gamma emitting substances are: iodine-131, cesium-137, cobalt-60, radium-226, and technetium-99m.[10] These materials are present in nature in solid, liquid or gaseous states.

The two chief sources of the aforementioned types of radiation in the environment are (i) radiation emitted from radionuclides naturally present in the earth's crest i.e. terrestrial source and (ii) radiation from radionuclides produced by high energy cosmic rays (particles) i.e. cosmic source. Before the discovery of cosmic rays in the 1920s, ionising radiation from radioactive elements in the earth's crust and radioactive gases like radon, generated from the decay of these radioactive elements, were believed to be the sole source of atmospheric radiation. The discovery of cosmic rays has revolutionised studies in the area of 'Ionising Radiation in Planetary and Interplanetary Space'.[11] Table 7.1 shows the cosmic ray dose rates at various altitudes and Table 7.2 shows the various ionising radiations emitted during the decay of radionuclides produced by cosmic rays as well as the terrestrial sources present in the atmosphere. The radiation exposure from cosmic rays increases with the increase in altitude. For people living in high altitude areas or frequent air-travellers, the average annual radiation exposure is therefore relatively higher.

Radiation exposure from natural sources takes place through both external as well as internal routes: (i) external radiation received from cosmic rays and terrestrial sources and (ii) internal dose from radionuclides taken into the body through ingestion of food materials etc., as well as from the inhalation of radon (222Rn), thoron (220Rn) gases generated from natural decay of uranium or thorium. The levels of natural radiation present in the atmosphere however, vary from place to place owing to differences in cosmic ray intensity with altitude, and in terrestrial radiation originating from soil and rocks.

Table 7.1[12]: Cosmic Ray Dose Rates at Various Altitudes

Elevation Above Sea level (m)	Equivalent DoseRate (μSv.y 1)	Elevation Above Sea level (m)	Equivalent Dose Rate (μSv.y 1)
0–150	260–270	1220–1828	390–520
150–305	270–280	1828–2438	520–740
305–610	280–310	1438–3408	740–1070
610–1220	310–390	> 3408	1070

Table 7.2[13]: Decay of Radionuclides produced by Cosmic Rays and Terrestrial Sources

Isotope produced by cosmic rays			Isotopes produced from terrestrial sources		
Isotope	Half life	RadiationEmitted	Isotope	Half life	Radiationemitted
^{14}C	5730 years	Beta	^{222}Rn(Radon)	3.82 days	Alpha
^{32}Si	650 years	Beta	^{218}Po(RaA)	3.05 months	Alpha
^{39}Ar	269 years	Beta	^{214}Pb(RaB)	26.8 months	Beta, gamma
^{3}H	12.3 years	Beta	^{214}Bi(RaC)	19.7 months	Alpha, beta, gamma
^{22}Na	2.6 years	Beta, gamma	^{210}Pb(RaD)	20.4 years	Beta
^{35}S	87 days	Beta	^{210}Bi(RaE)	5.0 days	Beta
^{7}Be	53 days	Gamma(EC)	^{210}Po(RaF)	138.4 days	Alpha
^{37}Ar	35 days	Gamma(EC)	^{220}Rn(Thoron)	55 secs	Alpha
^{33}P	25 days	Beta	^{216}Po(ThA)	0.158 secs	Alpha
^{32}P	14 days	Beta	^{212}Pb(ThB)	10.64 hours	Beta, gamma
^{24}Na	15 hours	Beta, gamma	^{212}Bi(ThC)	60.6 months	Alpha, Beta, gamma

Table 7.3[14]: Average Annual Radiation Exposure (mSv/yr)

Radiation		UNSCEAR	
Type	Source	World Avg.	Typical range
Natural	Inhalation	1.26	0.2-10.0
	Internal	0.29	0.2-1.0
	Terrestrial	0.48	0.3-1.0
	Cosmic	0.39	0.3-1.0
	Subtotal	2.40	1.0-13.0
Artificial	Medical	0.60	0.03-2.0
	Fallout	0.007	0-1±
	Others (including NPPs)	0.0052	0-20
	Subtotal	0.6	
Total		3.00	

Units of Dose
Unit Symbol Conversion Factors

Gray (SI)	Gy	1 J/kg = 100 rads
Rad	Rad	0.01 Gy = 100 erg/g
Sievert (SI)	Sv	1 J/kg = 100 rem
Rem	Rem	0.01 Sv

The studies conducted by the UNSCEAR have generated useful data on various natural as well man-made sources of radiation to understand the composition of the average radiation dose received by human populations, (see Table 7.3). The data shows that for most people on earth, naturally occurring background radiation is still the largest source of radiation exposure. There is considerable variation in the background levels of radiation around the world depending on geology and altitude and accordingly the radiation dose variation ranges between 1 and 10 mSv/yr, and at times can be more than 50 mSv/yr in some places.[15] In addition to these natural sources of radiation, increasing use of nuclear power for peaceful and military purposes has created additional sources of radiation exposure: such as fallout from weapons tests, radioactive releases from nuclear reactor operations and accidents, exposure due to radioactive waste disposal etc. The growing applications of radiation in the fields of medical therapies, food preservation, agriculture seed technology, manufacturing, and engineering etc have created additional sources of exposure for human population. These activities contribute to both *voluntary and non-voluntary* radiation exposures which can be compared with natural radiation levels.

The voluntary radiation exposures to human population can occur from various occupational activities, medical or research treatments, air travel etc; whereas involuntary radiation exposure may result from the transportation and disposal of radioactive waste, accidents in nuclear facilities, excessive exposure from orphan radiological substances etc. The non-voluntary exposures have been a matter of serious concern as they involve the risk of extremely high levels of radiation exposure due to unforeseen events such as nuclear or radiological accidents, malfunctioning of radiation therapy equipment or manhandling etc. The health impacts of voluntary as well as non-voluntary exposures have been extensively studied by the UNSCEAR and other agencies. These studies have mainly focused on understanding as to how much radiation increases the risk of a particular biological effect on health; what are the health effects that can occur at different levels of radiation exposure; and whether there is any exposure level that can be considered as safe. The next section discusses the various impacts of high level as well as low level radiation on human health.

Impact of Radiation on Human Health

Extensive research and studies have been conducted by various national as well as international agencies on the impact of radiation on human health. Much of the knowledge about risks from radiation is based on the study of more than 100,000 survivors of the atomic bombs at Hiroshima and Nagasaki, Japan. Other instances of industry workers and people exposed to large doses of radiation have also provided important knowledge on health effects. These studies show that the higher the radiation dose, the greater the chance of developing health risks including cancer and that there is a linear relationship between the dose received and the effect.[16] While, studying the severity and types of health effects from both voluntary as well non-voluntary exposures, the researchers have mainly focussed on the quantity and duration of the radiation exposure. Accordingly, the radiation exposures are divided into—high level radiation dose and low level radiation dose. While, the health impacts of high level radiation exposure have been well-established by scientific literature, the risks associated with low level radiation doses are still being debated and disputed by the scientists.[17] The biological effects of low level radiation exposure are so minor that health physicists have found it almost impossible to detect and clinically measure them. The health effects are apparent only at the higher levels of radiation exposure. In order to measure the biological effects at low levels of radiation exposure, the health physicists thus derive risks assumptions proportionate to the high levels of exposure (i.e. known effects of high level radiation from the survivors of Hiroshima and Nagasaki). This method of estimating risk is the 'linear no-threshold model (LNT).' It assumes that there is no 'threshold', which means there is no exposure level, below which, the risk is zero and any levels could be harmful.[18] It implies that the risk increases in proportion to the exposure, therefore, if exposure is doubled, the risk is also doubled and if exposure is reduced by half, the risk will also be reduced by half. This assumption is widely accepted by regulatory agencies all over the world for setting safe levels of occupational exposure for radiation workers in the nuclear fuel cycle facilities. Because the health risks are believed to be minimal at lower levels of radiation exposure; conservative levels of annual exposure limits to radiation workers are set. This assumption of a linear relationship between dose and effect however is challenged by certain health physicists, who argue that radiation exposure at a lower level can have a beneficial impact on the human body.[19] The next section discusses the impacts of both high level and low level radiation from voluntary and non-voluntary sources and their possible impacts as studied by the scientists through detailed epidemiological studies.

Effects of High Level Radiation Dose

It has been widely accepted by scientific groups that high doses of ionising radiation lead to observable harm in the individual that can be serious enough to be fatal. At higher doses up to 700 m/Sv, the body cells suffer irreversible damage and lose their natural strength to repair themselves or may either be changed permanently or die. While, the cells that die are of little consequence and gets replaced naturally, the cells that are changed permanently may go on to produce abnormal cells when they divide and these cells may then become cancerous.[20] At higher doses than 700 m/Sv, vital body parts begin to fail leading to death. There are two known immediate and long term risks associated with high levels of radiation exposure. The most common immediate or early effect of high level dose, as witnessed in Hiroshima-Nagasaki, Chernobyl, and other radiological accidents is the 'Acute Radiation Syndrome', whereas the long term risk manifests in various forms of cancer.[21]

Acute radiation syndrome (ARS) also known as 'radiation sickness' represents a constellation of health effects that present themselves within 24 hours of exposure to high amounts of ionising radiation. The high level radiation causes cellular degradation due to damage to DNA and other key molecular structures within the cells in various tissues; this destruction, particularly as it affects the ability of cells to divide normally, in turn causes serious symptoms. Relatively smaller doses result in gastrointestinal effects such as nausea and vomiting and symptoms related to falling blood counts such as infection and bleeding. Relatively larger doses can result in neurological effects and rapid death. For Acute Radiation Syndrome (ARS) to occur, the radiation doses must be large i.e., greater than 0.7 gray or 70 rads (700 mSv), caused by an external high energy X-rays, gamma rays, or neutrons.[22]

The three classic ARS symptoms are: (i) bone marrow syndrome (hematopoietic syndrome) usually occurs with a dose greater than approximately 0.7 Gy (70 rads) although mild symptoms may occur with a dose as low as 0.3 Gy or 30 rads, (ii) Gastrointestinal (GI) syndrome usually occurs with a dose higher than approximately 10 Gy (1000 rads) although some symptoms may occur at a dose as low as 6 Gy or 600 rads, and (iii) Cardiovascular (CV)/Central Nervous System (CNS) syndrome usually occurs with a dose greater than approximately 50 Gy (5000 rads) although some symptoms may occur at a dose as low as 20 Gy or 2000 rads.[23]

Cancer is considered to be the most significant long term risk associated with high level radiation in affected persons. Scientists have affirmed that the higher the radiation dose, the greater the chance of developing cancer. The chances of developing cancer increase with the increase in radiation doses. Cancer is generally defined as the uncontrolled growth of cells.[24] Ordinarily, natural processes control the rate at which cells grow and replace themselves. They also control the body's processes for repairing or replacing damaged tissue. Damage occurring at the cellular or molecular level, can disrupt the control processes, permitting the uncontrolled growth of cells cancer. This is why ionising radiation's ability to break the chemical bonds in atoms and molecules makes it such a potent carcinogen. Given the linear relationship between radiation doses and health impacts, high level radiation exposure is most likely to increase the risk of cancer in human body. The incidents of high level radiation causing 'Acute Radiation Syndrome' and cancer have been extensively analysed in studies conducted on the affected people from various radiation accidents.

Dose limits: Workers

For workers: According to ICRP recommendations, occupational exposure to radiation should not be higher than 50 mSv in any one year, and the annual average dose over five years must not exceed 20 mSv.

For the public: The dose limits for the general public are lower than those for workers. The ICRP recommends that the public should not be exposed to more than an average of 1 mSv per year.

Effect of Low Level Radiation Dose

The principal risk at lower levels of radiation exposure is an increase in radiation-induced cancer. There are also possibilities of hereditary diseases and other non-cancer effects.[25] Low level radiation is the major source of exposure for human population on earth. The health impacts of low level radiation are most difficult to measure. The UNSCEAR defines low radiation doses as 'those of 200 milligrays (mGy) or less and low dose rates as 0.1 mGy per minute (averaged over one hour or less) for radiations such as external X-rays and gamma rays.[26] Given the uncertainties of analysing the impacts of low level radiation, the UNSCEAR in its 56th session held in July 2008, declared a better understanding of the effects from low dose rate radiation exposure as its core priority for the period 2009–2013.[27] The UNSCEAR reported that "in general, increases in the incidence of health effects in populations cannot be attributed reliably to chronic exposure to radiation at levels that are typical of the global average background levels of radiation."[28] Furthermore, multiplying very low doses by large numbers of individuals does not give any meaningful results regarding health effects.[29]

The UNSCEAR has produced radiation risk estimates for cancer and hereditary effects arising from chronic low level radiation exposure. The UN committee has encountered difficulties in attributing specific cases of cancers to low-dose radiation exposure. These include: (a) the lack of specificity in the type or characteristics of cancer induced by radiation exposure; (b) the long delay (years or decades) between exposure and disease presentation; (c) the high spontaneous incidence of diseases associated with radiation in the ageing general population.[30] The UN committee has documented strong epidemiological evidence that exposure of humans to radiation at moderate and high levels can lead to the excess incidence of solid tumours in many body organs and leukaemia. However, the committee has noted that any increase in cancer incidence, believed to be caused by low-dose radiation exposures, is modest by comparison.[31]

Nuclear and Radiological Accidents

The first major event of the nuclear age, which caused widespread radioactive fallout and contamination, was the atomic bombing of Hiroshima and Nagasaki. The bombing exposed a large number of Japanese people to high and acute radiation doses. The studies conducted on the survivors of the bombing have provided important insights into the symptoms and dangers associated with radiation exposure. Since, the atomic bombing of Hiroshima and Nagasaki, several radiation incidents and accidents have taken place in different parts of world. These include accidents at industrial, medical, and military facilities, nuclear power plants and accidents due to orphan radiological sources, resulting in cases of ARS and cancer among affected individuals. During 1945-2007, the UNSCEAR has recorded a total of 203 accidents at facilities including nuclear weapon related installations, nuclear power plants, industrial facilities, orphan source accidents, accidents in academic/research work and accidents in medical use (See Table 7.4).

Table 7.4: Number of Accidents Resulting in early Acute Health Effects or Significant Population Exposures (Based on published information; excludes malicious acts and nuclear testing)

Type of accident	1945–1965	1966–1986	1987–2007	Sub-Total
Accidents at nuclear weapon facilities	17	5	2	24
Accidents at NPPs	2	7	2	11
Industrial accidents	2	50	28	80
Orphan source accidents	3	15	16	34
Accidents in academic/research work	2	16	4	22
Accidents in medical use[a]	Unknown	18	14	32
Total				203

[a]The IAEA [I25] and ICRP [I24] have reported more than 100 accidents in radiotherapy, however this information considers only serious radiation accidents in medicine.
Source: UNSCEAR, 2008 Report: Volume II, Annex C: Radiation Exposures in Accidents.

The accidents or incidents associated with sources of man-made radiation are classified on the basis of the International Nuclear and Radiological Event Scale (INES) designed by the International Atomic Energy Agency. On this scale events are rated from Level-1 (anomaly) to Level–7 (major accident) depending on the level of the radiological release and its impact.[32] The upper levels (4-7) are termed accidents and the lower levels (1-3) are termed incidents. Events with no safety significance are classified below scale at level 0 and are termed 'deviations'. In order to clearly communicate the safety significance of events to the public; a distinct phrase has been attributed to each level of INES. These are (from level 1 – level 7): 'anomaly', 'incident', 'serious incident', 'accident with local consequences', 'accident with wider consequences', 'serious accident' and 'major accident'. The 1986 accident at the Chernobyl nuclear power plant and 2011 accident at the Fukushima Dai-ichi nuclear station are designated at Level 7 on INES (See Fig. 7.1). These accidents were so rated due to their widespread impact on people and the environment.[33] According to UNSCEAR, the 203 accidents at different nuclear facilities during 1945-2007 resulted in a total of 147 deaths due to high level exposure, and 1246 early acute health effects (See Table 5). Of these 11 were serious accidents at nuclear power plants in different countries between 1945 and 2011 resulting in a total of 120 early effects and 34 deaths including the Chernobyl accident.[34]

Figure 7.1: INES Scale

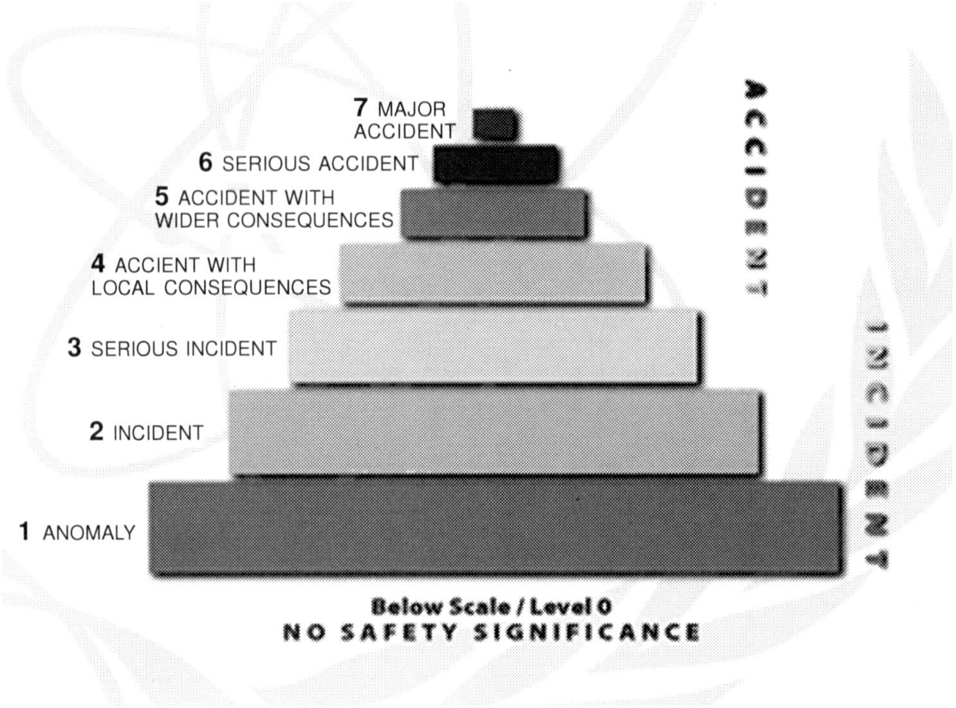

Table 7.5: Numbers of Deaths and early Acute Health Effects Due to Radiation Accidents

Based on published information; excludes malicious acts and nuclear testing

Type of accident	1945–1965	1966–1986	1987–2007	Total
Accidents at nuclear Weapons Facilities	36 early effects 12 deaths	10 early effects 3 deaths	1 early effects 1 deaths	47 early effects 16 deaths
Accidents at nuclear power plants	6 early effects 1 deaths	113 early effects 31 deaths	1 early effects 2 deaths	120 early effects 34 deaths
Industrial accidents	8 early effects 0 deaths	61 early effects 3 deaths	51 early effects 6 deaths	119 early effects 9 deaths
Orphan source accidents	5 early effects 7 deaths	98 early effects 19 deaths	205 early effects 16 deaths	308 early effects 42 deaths
Accidents in academic/ research work	2 early effects 0 deaths	22 early effects 0 deaths	5 early effects 0 deaths	29 early effects 0 deaths
Accidents in medical use	Unknown Unknown	470 early effects 4 deaths	153 early effects 42 deaths	623 early effects 46 deaths

Source: UNSCEAR, 2008 Report: Volume II – Annex C: Radiation Exposures in Accidents.

Chernobyl Accident exposed the people and the plant workers to large doses of radiation. The average effective dose to those persons most affected by the accident was assessed to be about 120 mSv for 530,000 recovery operation workers, 30 mSv for 115,000 evacuated persons and 9 mSv during the first two decades, after the accident, to those who continued to reside in contaminated areas.[35] The exposures however were much higher for those involved in mitigating the effects of the accident and those who resided nearby. The Chernobyl accident resulted in the release of a total of 5200 PBq of radioactivity (iodine-131 equivalent). Of the 600 workers present on the site on the day of accident, 134 received high doses (0.8-16 Gy) and suffered from Acute Radiation Syndrome (radiation sickness). Of these, 28 died in the first three months and another 19 died in 1987-2004 of various causes not necessarily associated with radiation exposure.

Table 7.6: Data for the 134 Patients Critically Injured in Chernobyl

Degree of ARS	Absorbed dose range (Gy)	Number of patients[a]	Number of early deaths[b]	Number of survivors
Mild (I)	0.8–2.1	41	0 (0%)	41
Moderate (II)	2.2–4.1	50	1 (2%)	49
Severe (III)	4.2–6.4	22	7 (32%)	15
Very severe (IV)	6.5–16	21	20 (95%)	1
Total	**0.8–16**	**134**	**28**	**106**

a Acute radiation syndrome was not confirmed in a further 103 workers.
b Percentage of treated patients in parentheses.
Source: UNSCEAR, 2008 Report: Volume II – Annex C: Early health Effects.

In addition, the majority of the 530,000 registered recovery operation workers received doses of between 0.02 Gy and 0.5 Gy between 1986 and 1990. UNSCEAR reported that, among the residents of Belarus, the Russian Federation and Ukraine, there had been up to the year 2005 more than 6,000 cases of thyroid cancer reported in children and adolescents who were exposed at the time of the accident, and more cases can be expected during the next decades. Many of those cancers were most likely caused by radiation exposures shortly after the accident. Apart from this increase, there is no evidence of a major public health impact attributable to radiation exposure two decades after the accident. There is no scientific evidence of increases in overall cancer incidence or mortality rates or in rates of non-malignant disorders that could be related to radiation exposure. The incidence of leukaemia in the general population, one of the main concerns owing to the shorter time expected between exposure and its occurrence compared with solid cancers, does not appear to be elevated. Although those most highly exposed individuals are at an increased risk of radiation-associated effects, the great majority of the population is not likely to experience serious health consequences as a result of radiation from the Chernobyl accident.[36]

The accident at the Daiichi nuclear power plant in Fukushima released about 940 PBq (iodine-131 equivalent) of radioactive material. In May 2013 UNSCEAR reported that "Radiation exposure following the nuclear accident at Fukushima Daiichi did not cause any immediate health effects. It is unlikely to be able to attribute any health effects in the future among the general public and the vast majority of workers."[37] The only exceptions were the 146 emergency workers who received radiation doses of over 100 mSv during the crisis. Thyroid doses in children were significantly lower than after the Chernobyl accident. Some 160,000 people were evacuated as a precautionary measure, and prolonged evacuation resulted in the deaths of about 1100 of them due to stress, and some due to disruption of medical and social welfare facilities. Certainly the main radiation exposure was to workers on site, and the 146 with doses over 100 mSv will be monitored closely for "potential late radiation-related health effects at an individual level." Six of them had received over 250 mSv—the limit set for emergency workers there, apparently due to inhaling iodine-131 fume early on. There was around 250 workers on site each day.[38]

Radiation Releases from Operation of Nuclear Plants in India

India's nuclear programme presently includes the full range of nuclear fuel cycle activities. During the operation of the various facilities involved in fuel cycle, the primary safety objective is to protect the plant personnel, the population in surrounding areas and the environment from radioactive contamination. In order to achieve this objective, the releases from nuclear fuel cycle facilities are monitored on a routine basis. Table 7.3 shows the constituents of the cumulative doses received by people all over the world. It shows that releases from nuclear plants contribute very little to the overall radiation dose received by people all over the world. Nevertheless, it is important to understand the contribution of radioactive releases from NPPs to the overall radiation dose received by people as well as NPP workers in India.

The radiation releases from India's nuclear fuel cycle facilities have been stringently regulated and monitored as per the AERB guidelines. Measurement of radiation levels and other pollutants has been an important objective of environmental radiation monitoring at all the nuclear fuel cycle facilities in India. For this purpose, Environmental Survey Laboratories (ESLs) have been established at each nuclear site to constantly monitor the radiation levels in and around these facilities. These laboratories collect the base-line environmental data in the pre-operational phase and thereafter carry out a systematic monitoring of radioactivity levels during operational stages of NPPs and other facilities and ensure that there is no build up of radioactivity in the environment, over and above the pre-operational natural levels.[39] The radiation monitoring is carried out through the vast collection of data on the size and distribution of the population, demographic and dietary patterns, and levels of activity in water, milk, soil, and food crops in the vicinity of the plant, which may contribute to human exposure.[40] During the operational stages of the NPP, radiation monitoring is carried out throughout the plant's operational lifetime to analyse the long term trend of radiological releases. The environmental radiation monitoring programmes at all the Indian nuclear sites have shown that operation of these facilities did not result in any measurable increase in the pre-operational or background radiation levels.

Safety Performance of Indian NPPs

India currently operates 21 nuclear reactors which comprise: Two BWR supplied by GE; Two Canadian built CNADU PHWRs; One Russian Designed LWR; and 16 indigenously built PHWRS. In the last four decades of the operation of NPPs, there has not been a single event at NPPs in India which has had an adverse radiological impact on the environment and the population. During 2007-11, a total of 155 events have been reported to the Atomic Energy Regulatory Board by the NPCIL. Of these, 142 were of Level-0 (i.e. no safety significance) as per the International Nuclear Event Scale (INES). The remaining 12 incidents were measured at Level-1 and one was of level-2 (see Table 9).[41] In the last five decades of operations of NPPs in India; there have been two major events of safety significance: the fire incident in the turbine building at Narora Atomic Power Station (1993, INES Scale Level-3) which resulted in a 17-hour total station blackout and the incident at Kakrapar Atomic Power Station caused by unintended power excursion (2004, INES Scale Level-2). Both these events had no radiological impact on the workers, the public or the environment.[42]

Table 7.7: INES Rating of Significant Events in NPPs during 2007-11

INES Levels	2007	2008	2009	2010	2011
0	28	22	23	33	36
1	8	2	0	1	1
2	0	0	0	0	1
3	0	0	0	0	0
>3	0	0	0	0	0
Total	**36**	**24**	**23**	**34**	**38**

Source: AERB Annual Report 2011.

Gamma Radiation Levels around the NPPs in India

In order to ensure that there is no increase in radiation background due to the operation of NPPs, the AERB has mandated both on-site as well as off-site monitoring to provide confirmatory evidence that the plant releases are within the prescribed limits.[43] The AERB also requires that the ESLs should have equipment and facilities for detection/analysis of very low levels of radioactivity in the environmental samples. In NPPs most of the radionuclides generated are beta or gamma radiation emitters. The radiation monitors are thus predominantly for the measurement of beta and gamma radiation.[44] The study on gamma radiation levels around the Kakrapar Atomic Power Station (KAPS) up to 30 km distance was carried out by using a Gamma Tracer (Genitron make) mounted on a vehicle. The gamma levels were measured for a period of one year. The study has shown that the hourly average of gamma radiation levels in all the routes of the vehicle on many occasions varied from 56 to 112 $nGyh^{-1}$ with an average of 77 $nGyh^{-1}$. The study has found this data to be comparable with the pre-operational data of nuclear station. The study shows the negligible impact of operation of KAPS on the surrounding environment.[45]

A comparative environmental parametric assessment for the Rawatbhata site based on the *Rapid* Environment Impact Assessment carried out for Unit 5-8 in September 1989 and the *Comprehensive* Environment Impact Assessment carried out for Unit 7-8 in May 2005 shows no significant increase at boundary post.[46] The total annual effective dose, from natural radiation sources, to the members of the Indian public is measured as 2.299 mSv/y; which can be compared with the global value of 2.455 mSv/y (Table 7.8). Table 7.9 gives the estimated natural background radiation levels in India major cities and Table 7.10 gives the estimated natural background radiation

Table 7.8[47]: Radiation Exposures to Indian Population from Different Natural Sources and its Comparison with the Reported Global Value

Radiation Sources	India		World	
	Annual Effective Dose (mSv/y)	Percent Contribution	Annual Effective Dose (mSv/y)	Percent Contribution
External				
Cosmic radiation	0.355	15.44	0.380	16.14
Terrestrial	0.379	16.48	0.480	19.55
Internal				
Cosmogenic nuclide (inhalation)	0.015	0.65	0.010	0.41
^{222}Rn and ^{220}Rn (inhalation)	1.235	53.72	1.275	51.94
Terrestrial	0.315	13.70	0.310	12.63
Total (Rounded Off)	2.30	100.0	2.5	100.0

Table 7.9[48]: Estimated Natural Background Radiation Levels in Indian Major Cities

City	Cosmic Ray component (mSv/y)	Terrestrial component (mSv/y)	Inhalation component (Rn +Tn) (mSv/y)	Cosmogenic component inhalation (mSv/y)	Ingestion through diet (mSv/y)	Total (mSv/y)
Bangalore	0.287	0.412	0.70	0.015	0.315	1.729
Chennai	0.287	0.536	0.55	0.015	0.315	1.703
Delhi	0.287	0.477	0.70	0.015	0.315	1.794
Hyderabad	0.287	0.875	1.09	0.015	0.315	2.582
Kolkatta	0.287	0.568	1.76	0.015	0.315	2.975
Mumbai	0.287	0.202	0.62	0.015	0.315	1.439
Nagpur	0.287	0.317	1.96	0.015	0.315	2.894
Trivandrum	0.287	0.412	0.70	0.015	0.315	1.729

Table 7.10[49]: Natural Background Radiation Levels at Some of DAE Installations in India (mSv/y)

Location	Background Radiation Dose (mSv/y)	Locations	Background Radiation Dose (mSv/y)
RAPS	0.60 – 1.01	NAPS	1.04 – 2.28
KAPS	0.63 – 0.91	UCIL, Jaduguda	0.99 – 2.22
TAPS	0.56 – 1.12	NFC, Hyderabad	1.63 – 2.79
BARC	0.45 – 0.70	MAPS	1.11 – 2.79
Kaiga	0.46 – 1.15		
IRE, Alwaye	0.56 – 1.12		

at some of the DAE installations in India (mSv/y). The environmental monitoring around the NPPs in India shows that radioactive releases into the environment are well within the limits prescribed by the Atomic Energy Regulatory Board. There are no instances of local people receiving unusually high levels of radiation dosages due to operation of NPPs as compared to the dose received from natural sources.

Radiation Dose Received by NPP Workers

In addition to the environmental radiation monitoring, the AERB guidelines on radiation protection require the institution of a 'comprehensive personnel monitoring programme' to measure both the external and internal dose received by individual workers at the nuclear installations.[50] The monitoring of personnel at NPPs is carried out by using a variety of monitors. For external dose measurement, devices like Thermo Luminescent Dosimeters (TLDs), Diamond Radiation Detectors (DRDs), extremity dosimeters, neutron dosimeters, etc are used; whereas for internal dose measurement techniques like bioassay, WB counting etc are prescribed by the AERB guidelines. Personal contamination monitors like friskers, hand/ foot monitors, and portal surveillance monitors are also used to measure the dose received by the person due to beta, gamma and neutron radiations.[51]

Table 7.11 shows the radiation dose received by number of workers in NPPs during the year 2011. During this year the AERB recorded four cases of exposure above 20 mSv. Table 7.12 shows the radiation doses received by workers in front end fuel cycle facilities in India. As per AERB guidelines, for an occupational worker the annual dose limit is 30-mSv, with the condition that it should not exceed 100 mSv over a span of five years.[52] The

Table 7.11: Radiation Doses Received by Workers in NPPs (2011)

NPP	Number of monitored persons	Avg. dose for monitored Persons (mSv)	No. of persons received dose	Average dose among dose receivers (mSv)	No. of workers received dose in the range		
					<20 (mSv)	20-30 (mSv)	>30 (mSv)
TAPS-1&2	1801	1.99	1468	2.44	1801	0	0
RAPS-1&2	999	1.12	670	1.67	999	0	0
MAPS-1&2	1332	1.51	1165	1.73	1332	0	0
NAPS-1&2	1266	1.02	834	1.55	1266	0	0
KAPS-1&2	1258	1.22	900	1.71	1254	1	3
KGS-1&2	1525	1.30	1172	1.70	1525	0	0
RAPS-3&4	1568	1.43	1191	1.88	1568	0	0
TAPS-3&4	1842	0.73	1192	1.13	1842	0	0
KGS-3&4	1379	0.92	816	1.56	1379	0	0
RAPS 5&6	1342	0.44	933	0.63	1342	0	0

Source: AERB Annual Report 2011.

specified annual limit for radiation exposure of temporary worker is 15 mSv. Individuals receiving excess radiation exposure are transferred to non-radiation facilities so as not to exceed the overall five year dose limit. Through such measures, the overall radiation exposure is controlled at an early stage so as to avoid breaching the AERB specified dose limits.

Table 7.12: Radiation Doses Received by Workers in Front End Fuel Cycle Facilities

Type of Facilities	Location	Number of Exposed Persons	Average Dose for Exposed Persons (mSv)	Maximum Dose of Exposed Persons (mSv)	Number of workers received dose in the range		
					<20 mSv	20–30 mSv	<30 mSv
Uranium mines (UCIL)	Jaduguda	698	4.07	9.39	698	0	0
	Bhatin	179	3.62	9.94	179	0	0
	Narwapahar	689	4.26	6.99	689	0	0
	Turamdih	462	5.73	10.42	462	0	0
	Bagjata	152	3.05	3.49	152	0	0
	Banduhurang	54	1.26	1.38	54	0	0
Uranium mill (UCIL)	Jaduguda	475	1.92	5.61	475	0	0
Thorium Mines and Mineral Separation	Chavara	53	0.42	1.41	53	0	0
	Chatrapur	161	0.94	7.74	161	0	0
	Manavalakurichi	187	5.75	14.11	187	0	0
Thorium mill (IREL)	Udyogamandal	293	4.56	16.9	293	0	0
Fuel fabrication (NFC)	Hyderabad	906	1.05	9.8	0	0	0

Source: AERB Annual Report 2011.

Epidemiological Studies on Impact of Low Level Radiation on NPP Workers

Epidemiology is widely defined as the study of the distribution of disease in populations and of the factors that affect this distribution. There are several radiation related epidemiological studies conducted by various agencies across the world. In India, the department of atomic energy has commissioned several epidemiological studies of NPP and uranium mine workers, in order to ensure that radiation exposures to workers at nuclear facilities and mines do not lead to any adverse health impacts. The studies on NPP workers have provided a useful distinction between radiation workers and non-radiation workers for analysing the health effects of low level exposure in nuclear facilities (See Table 7.13). Several parts of India such as state of Kerala, and some part of Tamilnadu where background radiation levels are naturally high are designated as high background radiation areas (HBRAs). Studies have been carried out on large numbers of people residing in these areas to study the biological impacts of high natural radiation. These studies have provided important data for comparative assessment of health effects on the local population and NPP workers and their families.

Table 7.13: Epidemiological Studies Conducted by Tata Memorial Centre (TMC)

NPP Site	Study published	Radiation Workers	Non-Radiation Workers	Total Worker	Spouse	Children	Total
Tarapur & BARC	March 1998	2155	1185	3340	2271	4478	10089
Kakrapara	March 1999	1214	200	1414	1022	1537	3973
Kaiga	December 2000	1228	325	1553	1162	1343	4058
Rawatbhata	June 1999	1009	2988	3997	3200	6486	13683
Narora	June 2000	932	666	1598	1433	3746	6777
Kalpakkam	March 2001	2445	3017	5462	3969	5589	15020

Source: Tata Memorial Centre Reports

In January 1992, the Nuclear Power Corporation of India Limited commissioned the Tata Memorial Cancer Centre (TMC) to analyse radiation impacts on NPCIL employees and their families. These studies were carried out at six power stations at Tarapur, Maharashtra; Rawatbhala, Rajasthan; Kalpakam, Tamilnadu; Narora, Uttar Pradesh; Kakrapar, Gujarat and Kaiga, Karnataka. One of the important results of these studies was that there was no increase in the prevalence of malignancies in the radiation workers as compared to non-radiation workers; nor was there any difference in the prevalence of malignancies in spouses and offspring of employees as compared to relevant control groups.[53] These studies have provided several useful indicators and generated reliable baseline data for carrying out further work.

During January and December 2006, the Health Physics Unit of BARC carried out a health survey of 598 villagers residing near Jaduguda. These included 152 males, 217 females and 229 children below the age of twelve.[54] As per the survey results in this group, the villages around the uranium mines, mills and tailings ponds have been surveyed from time to time by various agencies appointed by government and the reports have conclusively suggested that the incidence of diseases including cancer in these areas is not higher than the national average.[55] The survey specifically mentions that there is no detected case of congenital malformation and mental retardation; no cancer cases were found; the number of infertility cases amongst married woman were two, which is below the national average; and the villagers suffer from conventional health problems which could be seen in any village with similar socio-economic conditions.[56]

There are many areas in India with a much higher natural higher radiation exposure than the national average. In Kerala, people living along the high radiation areas receive 10 times the national average exposure. The regional Cancer Centre at Thiruvananthapuram thus conducted a study involving more than four lakh people, covering an area of 192 sq. km. In this study the entire population was surveyed in terms of their lifestyle, habits etc. and radiation levels—inside and outside the house—as well as in soil were measured. Cancer incidence was recorded from the population based cancer registry using internationally accepted methods. The overall cancer incidence was found to be the same between Karunagapally and Thiruvananthapuram although Karunagapally received a much higher radiation dose. Moreover, the incidence of cancer was found to be fifty per cent less than in other parts of India like the Northeast state of Mizoram where there were no nuclear power plants.[57] All the epidemiological studies conducted at Tarapur, Kakrapar, Kalpakkam, Rawatbhata, and Narora have suggested that cancer cases and birth defects among people living near nuclear installations conform to the national average.[59] There is clearly significant evidence to show that the nuclear fuel cycle in India has not increased radiation exposure, either for nuclear plant workers, or for the surrounding population.

Conclusion

The chapter concludes that natural sources of radiation by far are the largest sources of radiation exposure for the human population in India as well as in the world. Extensive monitoring of radiation levels in and around the NPPs in India has shown no significant increase of environmental radioactivity due to operation of NPPs. The increase in radioactivity levels in the environment matrices was registered only after the events of atmospheric nuclear explosions and the accident at Chernobyl, USSR in 1986. The radiation monitoring surveys clearly show that the doses received by the nuclear plant workers are significantly lower than the prescribed regulatory limits and pose no serious health risks to the workers and their families.

This is further confirmed by the epidemiological studies which examined the impact of radiation exposure on nuclear plant workers and their families. The epidemiological studies have concluded that the instances of cancer among the nuclear plant workers are comparable with the national cancer average. One can therefore surmise that the operation of nuclear plants and other fuel cycle facilities in India does not pose any serious risks to public health and environment. Although several civil society groups have expressed concerns over health impacts from operation of NPPs; this chapter shows that best practices have been adopted to eliminate any health and environment related risks to the population and environment.

NOTES & REFERENCES

1. After the Atomic Bombing of Hiroshima and Nagasaki, Prime Minister Jawaharlal Nehru asked Dr D S Kothari and Dr Homi Bhabha to study nuclear weapons and their impacts. The study was first published in 1956 and subsequently in 1958. See, Kothari D S (1959), "Nuclear Weapons and Their Effects", *Publications Division Govt of India*, New Delhi
2. Comprehensive test Ban Treaty Organization (CTBTO), "Nuclear Testing 1945 – Today", URL: http://www.ctbto.org/nuclear-testing/history-of-nuclear-testing/nuclear-testing-1945-today/, accessed on September 26, 2013. The efforts to promote nuclear test ban, culminated in the signing of Partial Test Ban Treaty (PTBT), by the US, the USSR and the UK on August 5, 1963 in Moscow. It was heralded as an important step in the path towards achieving general and complete disarmament. The PTBT was also regarded an important step in safeguarding the health of living beings from the highly hazardous radioactive fallout from atmospheric nuclear weapon tests.
3. United Nations Scientific Committee on Effects of Atomic Radiation (UNSCEAR), "Historical Milestones", URL: http://www.unscear.org/unscear/about_us/history.html, accessed on September 27, 2013.
4. The International Commission on Radiological Protection (ICRP) and The International Commission on Radiological Units and Measurements (ICRU) were established much before the UNSCEAR. Both the ICRP and the ICRU are non-governmental academia, and do not act in the name of or on behalf of governments. The NRPB and BEIR are the national agencies in US and UK respectively and have published several studies on radiation and their effects.
5. Committee on the Biological Effects of Ionising Radiation (BEIR) is a committee of the National Research Council of the USA which publishes reports for the US government on the effects of ionising radiation.
6. World Health Organization, "What is Ionizing Radiation?", URL: http://www.who.int/ionizing_radiation/about/what_is_ir/en/, accessed on September 20, 2013
7. For more discussion on radiation, see United Nations Scientific Committee of Effects of Atomic Radiation (UNSCEAR).
8. Health Physics Society, "What Types of Radiation Are There?", URL: http://www.hps.org/publicinformation/ate/faqs/radiationtypes.html, accessed on September 25, 2013.
9. Ibid, *Types of Radiation.*
10. Ibid, *Types of Radiation.*
11. Several scientific studies conducted in 1940s and 50s show the presence of radiation from

cosmic rays and the constant decay of various cosmogenic particles. While the presence of cosmic rays was an important scientific discovery, the origins of cosmic rays continue to be debated by the scientists. The presence of radiation has raised concerns regarding in the safety of crews during space travel. Most recently, presence of radiation has become a major obstacle to the possibility of first man-mission to Mars.

12. Ramachandran, T.V. (2011), "Background radiation, people and the environment", *International Journal of Radiation Research*, 9(2): 63-76.
13. Ibid, *Ramachandran, T.V.*
14. UNSCEAR (2008), "Report to the General Assembly, Annex B: Exposures of the public and workers from various sources of radiation", URL: http://www.unscear.org/docs/reports/2008/09-86753_Report_2008_Annex_B.pdf, accessed on September 28, 2013
15. Nuclear Regulatory Commission, "Fact Sheet on Biological Effects of Radiation", URL: http://www.nrc.gov/reading-rm/doc-collections/fact-sheets/bio-effects-radiation.html, accessed on September 30, 2013
16. Nussbaum RH., (1998), "The linear no-threshold dose-effect relation: is it relevant to radiation protection regulation?", *Medical Physics,* 25(3):291-9.
17. Environment Protection Agency (US-EPA), "Health effects of radiation, United States", URL: http://www.epa.gov/radiation/understand/health_effects.html, accessed on October 02, 2013
18. Environment Protection Agency (US-EPA), "Estimating Risk", URL: http://www.epa.gov/radiation/understand/risk.html, accessed on October 02, 2013
19. T.D. Luckey (2008), "Radiation Hormesis: The Good, the Bad, and the Ugly", *Dose Response,* 2006; 4(3): 169–190, URL: http://www.ncbi.nlm.nih.gov/pmc/articles/PMC2477686, accessed on October 02, 2013. For more discussion on late impacts of low level radiation exposure, see Ivanov V K. (2007), "Late cancer and non-cancer risks among Chernobyl emergency workers of Russia", *Health Physics*, Vol. 93, pp. 470–79, Noshchenko A.G., Moysich K.B., Bondar A., Zamostyan P.V., Drosdova V.D., Michalek A. M. (2001), "Patterns of acute leukaemia occurrence among children in the Chernobyl region", *International Journal of Epidemiology*, Vol. 30, pp. 125-29, and Williams D., Baverstock K. (2006), "Too soon for a final diagnosis", *Nature*, 440:993–94. The dominant view about the low level radiation is that at very low levels of radiation exposure, it is almost impossible to conclusively observe effects of low level of radiation, due to low probability and confounding factors. LNT is thus taken as "to be on the safe side hypothesis". The number of cancer incidents assumed by this hypothesis is never proved in epidemiological surveys. The assumed value of cancer risk and the lack of actual evidence for such value is the source of the confusion. On the other hand, the proponents of radiation hormesis argue that low levels of radiation have beneficial effects in reducing cancer and genetic effects.
20. International Atomic Energy Agency, "Radiation Biology: A Handbook for Teachers and Students", URL: http://www-pub.iaea.org/MTCD/publications/PDF/TCS-42_web.pdf, accessed on October 02, 2013.
21. Environment Protection Agency (US-EPA), "Radiation Protection: Health Effects", URL: http://www.epa.gov/radiation/understand/health_effects.html, accessed on October 03, 2013.
22. Centre for Disease Control and Prevention, "Fact Sheet: Acute Radiation Syndrome: A Fact Sheet for Clinicians", URL: http://www.bt.cdc.gov/radiation/arsphysicianfactsheet.asp, accessed on October 03, 2013.
23. Ibid, *Fact Sheet: Acute Radiation Syndrome: A Fact Sheet for Clinicians.*
24. National Cancer Institute, "*Defining Cancer*", URL: http://www.cancer.gov/cancertopics/cancerlibrary/what-is-cancer, accessed on October 04, 2013
25. UNSCEAR (2000), "Annex G: Biological effects at low radiation doses", URL: http://www.unscear.org/docs/reports/annexg.pdf, accessed on October 04, 2013.

26. Report of the United Nations Scientific Committee on the Effects of Atomic Radiation 2010, Fifty-seventh session, includes Scientific Report: summary of low-dose radiation effects on health, URL: http://www.unscear.org/docs/reports/2010/UNSCEAR_2010_Report_M.pdf, accessed on October 04, 2013.
27. UNSCEAR (2012), "Biological Mechanisms of Radiation Actions at Low Doses: A white paper to guide the Scientific Committee's future programme of work", p.1, URL: http://www.unscear.org/docs/reports/Biological_mechanisms_WP_12-57831.pdf, accessed on October 05, 2013.
28. Ibid, p.2-3
29. UNSCEAR (2006), "Epidemiological Studies on Radiation and Cancer", URL: http://www.unscear.org/docs/reports/2006/07-82087_Report_Annex_A_2006_Web_corr.pdf, accessed on October 06, 2013.
30. Report of the United Nations Scientific Committee on the Effects of Atomic Radiation 2010, Fifty-seventh session, URL: http://www.unscear.org/docs/reports/2010/UNSCEAR_2010_ Report_M.pdf, accessed on October 06, 2013
31. Ibid.
32. International Atomic Energy Agency, "INES: The International Nuclear and Radiological Event Scale", URL: http://www.iaea.org/Publications/Factsheets/English/ines.pdf, accessed on October 06, 2013.
33. International Atomic Energy Agency (2008), "INES: The International Nuclear and Radiological Event Scale: A User's Manuel", URL: http://www-pub.iaea.org/MTCD/publications/PDF/INES2009_web.pdf, accessed on October 10, 2013.
34. This does not include the Fukushima accident. Including Fukushima accident there will be total 12 accidents at nuclear power plants facilities.
35. UNSCEAR, "The Chernobyl accident: UNSCEAR's assessments of the radiation effects", URL: http://www.unscear.org/unscear/en/chernobyl.html, accessed on October 5, 2013.
36. UNSCEAR Chernobyl Repot.
37. UNSCEAR, "The Fukushima-Daiichi nuclear power plant accident: UNSCEAR's assessment of levels and effects of radiation exposure due to the nuclear accident after the 2011 great east-Japan earthquake and tsunami", URL: http://www.unscear.org/unscear/en/fukushima.html, accessed on October 15, 2013.
38. Ibid.
39. Bhabha Atomic Research Centre, "Environment Radiation Monitoring", URL: http://www.barc.gov.in/randd/erm.html, accessed on October 04, 2013.
40. Ibid.
41. AERB (2011), "Atomic Energy Regulatory Board: Annual Report",
42. Chande, S.K. (2011), "Addressing Safety of Indian Nuclear Power Plants", *Atomic Energy Regulatory Board,* URL: http://www.aerb.gov.in/AERBPortal/pages/English/t/documents/SAFETYOFINDIANNPPs.pdf, accessed on October 15, 2013.
43. AERB Safety Guide NO - AERB/NPP-PHWR/SG/D-12, "*Radiation Protection Aspects in Design for Pressurized Heavy Water Reactor Based Nuclear Power Plants*", October 2005, p.31, URL: http: //www.aerb.gov.in/AERBPortal/pages/English/t/publications/CODESGUIDES/D-12.PDF, accessed on October 15, 2013.
44. Ibid, p.32.
45. S. Venkataraman et.al. (2010), "Monitoring of Gamma Radiation Levels Around Kakrapar Atomic Power Station", *Indian Association for Radiation and Protection*, Vol, 33 (4), pp.171-173.
46. The rapid EIA carried out at RAPS for Unit 5-8 in Sept 1989 shows the radiation levels 2.18 mrem/year whereas the radiation levels measured. during the comprehensive EIA for Unit 7 & 8 shows 3.04 mrem/year at the boundary wall as per information obtained from

the NPCIL authorities.

47. See *Ramachandran 2011,* p.67.
48. See *Ramachandran 2011,* p.68.
49. See *Ramachandran 2011,* p.68.
50. Ibid, *AERB safety guide "Radiation Protection during Operation of NPPs" (AERB/SG/O-5).*
51. Ibid, *AERB safety guide "Radiation Protection during Operation of NPPs" (AERB/SG/O-5).*
52. AERB Annual Report (2011), "Environmental Safety and Occupational Exposures", Government of India, pp. 58-65.
53. Tata Memorial Centre (1998), "Effect of Low-Dose Ionizing Radiation Among the Employees at the Tarapur DAE Centre: A Cross-Sectional Study", *The Epidemiological Studies Cell, Tata Memorial Hospital, Mumbai,* p.1-74.
54. Lok Sabha Unstarred Question No. 3331, "Ill Effects of Radioactive Waste", *Government of India, Department of Atomic Energy,* URL: http://dae.nic.in/writereaddata/3331_lsus160311.pdf, accessed on September 30, 2013.
55. Ibid, *Lok Sabha Unstarred Question No. 3331.*
56. Ibid, *Lok Sabha Unstarred Question No. 3331.*
57. Pramesh, C.S. & Badwe Rajendra (2012), "Nuclear Radiation and Health—the True Story", *Nuclear India,* Vol.47 (1), Jan-Mar 2012, p. 47.
58. Ibid, p.48.

CHAPTER 8

Nuclear Emergency Preparedness in India

Kapil Patil

Introduction

Nuclear power plants are designed, constructed, and commissioned with utmost care, so that operating personnel, public in the vicinity and the environment are protected from risks of undue radiation exposure. Operational safety has been the prime concern of the designers and operators of the nuclear power plants. All systems and components in the NPPs are designed to be 'fail safe', and premised on 'defence-in-depth' principle, wherein multiple layers of safety systems are provided to achieve maximum safety. However, despite the inclusion of layered safety systems; emergency plans are drawn up to deal with any eventuality involving large-scale radioactive fallout. A comprehensive off-site emergency plan is essential to mitigate the consequences of a nuclear accident, if any and to minimise radiation exposure to the public. The off-site nuclear emergency management that comprises of various technical, administrative and regulatory arrangements has evolved over the years, drawing upon lessons from past accidents like Three Mile Island and Chernobyl.

The recent Fukushima Dai-ichi accident in Japan revealed the need for further strengthening of emergency preparedness programmes in all nuclear power producing countries. It highlighted the need for enhanced inter-agency co-ordination and improved auxiliary infrastructure in the form of monitoring systems, transport arrangements, de-contamination facilities, etc. In India, members of certain civil society groups have raised numerous concerns about the technical as well as administrative aspects of nuclear emergency preparedness in the country.[1] The question whether Indian authorities are prepared to mitigate the consequences of large-scale radioactive fallout, has triggered a debate about the state of India's nuclear emergency preparedness programme.

In this backdrop, the chapter is primarily concerned with the off-site emergency preparedness plans and capabilities developed in India by the Department of Atomic Energy (DAE) and the National Disaster Management Authority (NDMA) which would be crucial for the effective management and mitigation during the off-site nuclear emergency.[2] It also seeks to understand how nuclear and civilian agencies in India have implemented the lessons of the Fukushima crisis especially with respect to critical capacity building in the form of advances technological systems, new organisational structures and the development of infrastructure in areas around the NPPs. The Fukushima crisis offers unique lessons for off-site emergency management and the implementation of these lessons would significantly enhance India's capability to handle off-site nuclear emergencies, while minimising the social and environmental consequences of any accident that may occur in future.

Nuclear Emergency Management

The protection of life and the environment from the consequences of both natural as well as man-made disasters constitutes an important task of socio-economic planning in modern societies. The consequences from certain natural disasters like earthquakes, floods, hurricanes, wildfires, etc. have been traditionally well known. Emergency preparedness plans and procedures for managing such natural disasters have also been fairly well-developed. In case of natural environmental imbalances like floods after excessive rains, approaching storms, earthquakes etc; the information about the disaster can be gathered through 'direct sensing' by the population.[3] People can estimate the imminence and magnitude of danger in such situations and decide upon possible self-protective measures such as moving to safer places, sheltering, appropriate prophylaxes, etc.

However, an accident at nuclear power installations poses unique challenge in managing and mitigating the subsequent consequences in the public domain. In case of a nuclear accident, the population in the surrounding areas would fail to take measures to avoid exposure to high dosage of radiation. The information regarding the magnitude of the accident (core damage) and stack releases can be made available only by a group of technical experts at the nuclear power plant. Therefore, indirect or remote sensing replaces direct sensory observation as a source of information about the evolving nuclear emergency at the nuclear power station.[4] While humans can perceive and analyse the progress of natural disasters, only technical experts can report the progress of nuclear disaster.

Information Availability during Natural Disasters

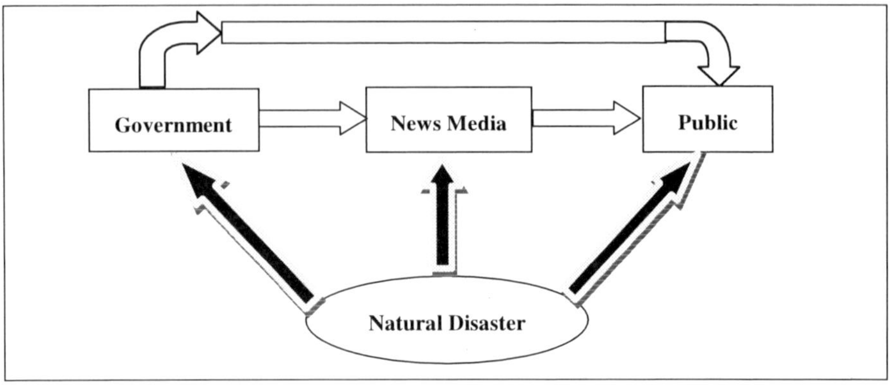

Primary sources of information; Secondary sources of information.
Source: Zeigler, Johnson and Brunn, 1983[5].

In case of a nuclear emergency, the agent causing disaster is ionising radiation which cannot be seen or felt by the human senses. The radiation field can be measured only through radiation monitoring devices if one can overcome the calibration problems that may occur. The public is therefore, fully dependent on a central authority such as the utility company or the regulatory agency to provide information about the possible self-protection measures to be taken during the nuclear accident.

Information Availability during Nuclear Disasters

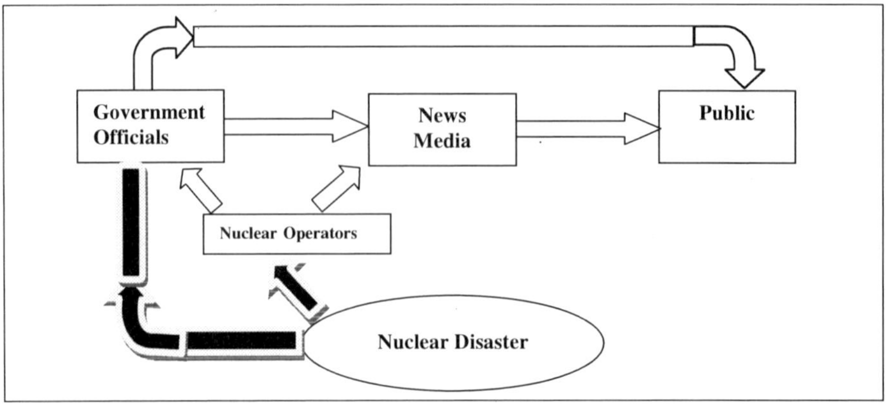

Primary sources of information; Secondary sources of information.
Source: Zeigler, Johnson and Brunn, 1983.

It is important therefore, to clearly lay down procedures for declaring an off-site emergency in the public domain through well-established regulatory framework. The legal and regulatory framework should also establish procedures for nuclear operators to develop comprehensive off-site emergency preparedness plans for mitigating the consequences of radioactive fallout in the public domain. To protect people from excessive radioactive exposure during nuclear emergencies, the nuclear authorities require diverse technical systems such as radiation monitoring networks, geographical information systems and decision support systems to decide upon wide range of radiation countermeasures.

A nuclear emergency cannot be managed by the nuclear authorities alone; the role of local and regional civic administration is crucial for implementing radiation counter-measures like mass evacuation during the off-site emergency. Carrying out mass evacuation with increased background environmental radiation levels in certain areas could be exceedingly difficult, if the civic agencies are not prepared to handle such emergencies. The civic administration must therefore, be involved in an off-site nuclear emergency planning and acquire the requisite capabilities to implement radiation counter-measures. Additionally, the development of infrastructure in the form of widespread roads, transport arrangements, health services, etc in areas around NPPs is also of paramount importance. Given the importance of technological systems, public infrastructure and organisational capabilities for off-site nuclear emergency management, the following sections discuss the regulatory, technical and organisational aspects of off-site emergency preparedness at the NPPs in India.

Legal and Regulatory Framework for Emergency Preparedness in India

Since the development of nuclear power for commercial purposes, the world has witnessed three major accidents in the history of nuclear power: Three Mile Island, USA in 1979, Chernobyl, Ukraine in 1986 and more recently Fukushima Dai-Ichi, Japan in 2011. Till the time of the TMI accident, radiation risks to the public were assumed to be very low and the debate over nuclear power safety was largely in terms of hypothetical accident scenarios. The radiological consequences of any accident that may occur were believed to affect only low population zones extending 2 to 3 miles from nuclear plant sites.[6] In 1977, the Nuclear Regulatory Commission (NRC)

USA, published a *Regulatory Guide 1.101* which specified more clearly the requirements for drafting emergency plans.[7]

In 1978, the US-NRC report on, "Planning Basis for the Development of State and Local Government Radiological Emergency Response Plans in Support of Light Water Nuclear Power Plants (NUREG-0396)" concluded that a "spectrum of accidents (not the source term from a single accident sequence) should be considered in developing a basis for emergency planning."[8] Additionally, NUREG-0396 introduced the concept of two Emergency Planning Zones (EPZs); the short-term 10-mile plume exposure pathway and the long-term 50-mile ingestion exposure pathway.[9]

The Three Mile Island accident in Pennsylvania, USA on 28th March 1979, transformed the idea of emergency planning zones into a mandatory licensing requirement. Evacuation planning within 10 miles of a plant site became a pre-requisite for licensing in the USA. The 10 mile radius was selected because in a core-melt accident 'the probability of large doses of radiation drops off substantially at about 10 miles from the reactor."[10] Subsequently in 1986, the NRC amended rule 10CFR50.47(b)(10) of the emergency guidelines, requiring that states with a population within the 10-mile emergency planning zone of NPP consider including potassium iodide (KI) as a protective measure for the general public to supplement sheltering and evacuation in the event of a severe accident.[11]

The Chernobyl accident in 1986 however, highlighted for the first time, the challenges of handling large-scale radioactive fallout and its trans-boundary consequences. The national and international emergency preparedness guidelines at the time of Chernobyl accident were far from adequately complete, in respect of radiation counter-measures and emergency planning for handling large-scale accidental radioactive fallout. The Chernobyl accident established the need for developing decision support systems, scientific codes, guides, meteorological and topographical databases, etc. for measurement of radioactive contamination in the environment and the consequent risks to human health.[12] The regulatory interventions after Chernobyl accident therefore, focussed primarily on developing new safety codes for NPPs and decision support tools as part of off-site emergency planning in most of the European countries. The new safety codes and support systems attempted to provide for a more coherent off-site emergency guidance to any future events at the NPPs. The new radiation monitoring systems and decision support tools were subsequently adopted by many nuclear power producing countries outside Europe.

The nuclear emergency preparedness programme in India has evolved over the years deriving numerous lessons from TMI and Chernobyl accidents. In India, the Atomic Energy Regulatory Board (AERB) is responsible for issuing guidelines for nuclear emergency preparedness. The AERB published the first set of off-site emergency preparedness guidelines in October 1999, and subsequently revised them in March 2000.[13] A comprehensive emergency preparedness plan is a mandatory licensing requirement for operation of NPPs in India. The AERB has to ensure the development of emergency response plans by the nuclear operators in conformity with international practices so that any potential eventuality of undue radiological risks for plant, personnel and the public, is handled effectively.[14] The AERB has formulated various safety codes and guidelines which specify the necessary requirements, with respect to the emergency preparedness at the Indian NPPs.

(i) AERB Safety Code "Regulation of Nuclear and Radiation Facilities" (AERB/SC/G, 2000) stipulates the minimum safety related requirements, including that for emergency preparedness, to be met by a nuclear or radiation facility to qualify for regulatory consent at every stage. Prior to issuance of licence for operation of a NPP, the AERB ensures that the approved emergency preparedness plans are in place and tested.[15]

(ii) The Safety Code "Safety in Nuclear Power Plant Operation" (AERB/SC/O, 2008) sets down the requirement for development of an emergency preparedness plan and for a high degree of emergency preparedness on the part of the plant management. The emergency preparedness programme ensures that in the event of an emergency situation, appropriate measures will be taken to mitigate the consequences.[16]

(iii) AERB Safety Guide "Role of the Regulatory Body with Respect to Emergency Response and Preparedness at Nuclear and Radiation Facilities" (AERB/SG/G-5, 2000) prescribes the role of the regulatory body during emergencies at nuclear and radiation facilities. It provides the information necessary to assist the facilities, and other participating/collaborating agencies, to fulfil the requirements stipulated in the safety code.[17]

(iv) The safety guide "Intervention Levels and Derived Intervention levels for Off-Site Radiation Emergency" (AERB/SG/HS-1, 1993) lays down the intervention levels (ILs) and derived intervention levels (DILs) for initiating countermeasures in the public domain following a nuclear accident or radiological emergency.[18]

(v) Safety Guide "Preparedness of the Plant Management for Handling Emergencies at NPPs" (AERB/SG/O-6, 2000) supplements the Code of Practice on Safety in NPP Operation. It provides the important guidelines relevant to the preparation and implementation of emergency response plans by the Plant Management.[19]

(vi) The regulatory requirements for preparing and maintaining an emergency response plan for off-site emergency are given in the safety guide "Preparation of Off-Site Emergency Plans for Nuclear Installation" (AERB/SG/EP-2, 1999).[20]

(vii) The safety manual on "Radiation Protection in Nuclear Facilities" (AERB/NF/SM/O-2, Rev. 4, 2005), issued under the Code of Practice on Safety in Nuclear Power Plant Operation, provides the necessary information for the facilities to carry out radiation protection functions including emergency preparedness.[21]

In addition to the AERB safety guidelines, the Indian Parliament passed the 'Disaster Management Act' in 2005. The Act provides for effective management of disasters including accidents at nuclear power plants (NPP) in India.[22] As mandated by the Act, the National Disaster Management Authority (NDMA) was set up in 2006, with the prime minister of India as the chairperson and the State Disaster Management Authority (SDMA) at the state levels, with chief ministers as the chairpersons. The NDMA has published national guidelines on nuclear and radiological emergencies in 2009 which supplements the existing nuclear/radiological emergency management framework.[23] The NDMA national guidelines require developing 'disaster management plans' at all levels of administration—state, district and local. All the states/union Territories hosting NPPs need to prepare disaster management plan in consultation with the nuclear plant authorities to ensure effective preparedness during off-site nuclear emergencies. These plans include measures to be taken by the concerned agencies for prevention of disasters and for the mitigation of their effects.[24]

The Declaration of Off-Site Emergency and the Implementation of Radiation Counter-measures

The AERB safety guides identify four types of emergencies at NPPs depending on their severity: (i) plant emergency alert, (ii) plant emergency, (iii) site emergency, and (iv) off-site emergency. Nuclear power plants in India are operated by the Nuclear Power Corporation of India Limited (NPCIL), which is primarily responsible for handling first three types of emergencies. The off-site emergency, which assumes the highly unlikely possibility of large-scale accidental radioactive releases into the public domain, is a responsibility of the district administration with technical support and guidance received from the DAE. The planning and preparedness for off-site emergency is carried out in a 16 km radius around the plant designated as the 'Off Site Emergency Planning Zone'.

Table 8.1: Types of Emergency

Type of emergency	Affected zones	Responsible agency
Emergency standby	Stack location	Plant Management
Personnel emergency	Plant area	Plant Management
Plant emergency	Plant area	Plant Management
On-site emergency	Plant area including Exclusion zone (1.6 km)	Plant Management
Off-site emergency	Emergency planning zone (16 km)	Plant Management, District Admin, State Govt. & NDMA

Source: 'AERB Safety Guide no. AERB/SG/O6 titled 'Preparedness of the operating organisation for handling emergencies at nuclear power plants'.

In the operational history of NPPs in India, there has not been any off-site emergency causing adverse radiological impact on the environment.[25] The NPCIL is required to report any safety event of major or minor significance to the AERB as part of the regulatory norms for NPPs. The AERB approves the on-site as well as off-site emergency plans of NPCIL which are drawn separately for each NPP. The NPCIL and DAE are required to routinely review and update the emergency plans and procedures for off-site emergency preparedness to meet the requirements of Convention on Nuclear Safety[26] which obliges the participating states to ensure nuclear emergency preparedness at the NPPs as per the following provisions:

(i) Each Contracting Party shall take the appropriate steps to ensure that there are on-site and off-site emergency plans that are routinely tested for nuclear installations and cover the activities to be carried out in the event of an emergency. For any new nuclear installation, such plans shall be prepared and tested before it commences operation above a low power level agreed by the regulatory body.

(ii) Each Contracting Party shall take the appropriate steps to ensure that, insofar as they are likely to be affected by a radiological emergency, its own population and the competent authorities of the States in the vicinity of the nuclear installation are provided with appropriate information for emergency planning and response.

Declaration of Off-site Emergency

While the NPPs in India are operated under a comprehensive regulatory oversight; in the unlikely event of off-site emergency, the NPCIL has to initiate the emergency measures and inform the public about the steps to be taken to avoid health hazards of radioactive contamination. The AERB safety guide "Preparation of Off-Site Emergency Plans for Nuclear Installation" (AERB/SG/EP-2, 1999 specifies the criteria for declaring and terminating the off-site emergency.[27] The safety guide also specifies various measures to protect people from radiation exposure during the nuclear accident. Depending upon the nature and duration of an accident, the emergency would evolve in the following stages:

Stage 1: This involves the declaration of a 'site emergency' by the site emergency director (SED) at the NPP. The trained staff at the station forms a site emergency response organisation under the command of the 'Station Director' to prevent the escalation of emergency to a higher level. If an emergency underway at the station escalates to the stage of dangerously high levels of radiation at the site boundary; the station director alerts the Off-site Emergency Director (OED) for activating off-site emergency plans and initiating the emergency response measures in the public domain. As per the AERB guidelines, the off-site emergency is declared when:

(i) Gamma radiation levels at 1m above the ground in off-site locations exceed 0.001mSv/h as stipulated in Derived Intervention Levels (DIL).[28] (The guidelines provide detailed Derived Intervention Levels (DIL) for radiation countermeasures)

Stage 2: On receiving the off-site emergency alert at the NPP, the Crisis Management Group within the DAE is immediately activated to co-ordinate off-site emergency measures. Also the National Crisis Management

Committee (NCMC) under the chairmanship of the cabinet secretary (government of India) is convened to review the off-site emergency measures. The CMG assesses the need for radiation countermeasures and provides expert advice to the Off-site Emergency Committee under the command of district commissioner. The CMG provides technical advice to the plant authorities and district administration for initiating radiation countermeasures in the affected areas. The AERB guidelines suggest the initiation of radiation counter-measures when:

(i) Levels of contamination in water and food materials in off-site locations exceed the stipulated DIL.[29] (The guidelines provide detailed Derived Intervention Levels (DIL) for radiation countermeasures)

Stage 3: This involves the Off-site Emergency Director (OED)—the district commissioner—who, after receiving guidance from the CMG and NCMC, will give authoritative advice to police, transport, health, and other off-site agencies to carry out evacuation, if necessary, to protect public from radioactive exposure. The NPCIL Technical Adviser and Emergency Response Teams (ERTs) will liaise with the OED in this task. The CMG coordinator will be the principal government spokesperson for briefing the news media.

Implementation of Radiation Counter-measures

An unlikely scenario of major nuclear accident could lead to high airborne activity and deposition of radioactive substances over large areas on the ground as witnessed in Chernobyl and Fukushima accidents. Spread of radioactivity in such a situation will depend on various parameters like the source term, release height, meteorological conditions, topography around the site, etc.[30] The radioactive materials released from the NPPs disperse in the form of a plume or cloud and the concentration of radioactivity in the plume decreases with distance from the site, being diluted and dispersed as it moves downwind.[31] The dispersion of such materials in the atmosphere through wind and it's possible in-take into the human body through water, food and other means could have wide-ranging consequences on human health. Generally, radiation exposure to the public, which can be external and/or internal, is envisaged through the following pathways:

- Immersion in plume-direct on ground and above ground
- Deposition of activity on walls and ground (contamination of surfaces and areas)

- Intake through ingestion route
- Intake through cuts and wounds
- Intake through grass, cow, milk pathway
- Intake through food chain and water[32]

The exposure to radiation primarily creates two types of biological effects: stochastic effect and deterministic effect. In the stochastic effect, the chance or the probability of the effect occurring, is statistical in nature and is a function of dose without any threshold. The known examples are cancer and genetic effects.[33] For deterministic effects, groups of cells or tissues are extensively damaged due to exposure to very high radiation doses and biological effects.[34] The general objective of implementing radiation countermeasures is to reduce health consequences; in particular, to avoid deterministic effects and to keep stochastic effects as low as possible. The AERB guidelines envisage following radiation countermeasures to be implemented during the off-site emergency:

(a) Sheltering: Sheltering is to be implemented as a positive assured countermeasure and people in the affected sectors would be directly advised of the need to stay indoors, close doors and windows, and take other simple protective measures. Sheltering is advised when external gamma radiation dose measures between 5mSv-20 mSv persistently for 10 hours.[35]

(b) Administering of potassium iodate tablets: Members of the public who might be exposed during the passage of the radiation plume will be provided with potassium iodate tablets to minimise the effects of any uptake of radioactive iodine. This countermeasure is implemented when effective radiation dose ranges between 50mGy-500mGy to the Thyroid.[36] Regarding the iodine prophylaxis, the nuclear plant authorities rely on the distribution of iodine tablets through a central storage at the time of emergency.

(c) Evacuation: Evacuating the public from the affected areas would reduce the risk of prolonged exposure to the radioactive materials released from the NPP during an accident. The evacuation is carried out when the effective radiation level measures between 20mSv-100mSv for more than 10 hours.[37] Internationally, an estimated or anticipated effective dose of 30 to 500 mSv is taken as the intervention level for the implementation of an evacuation.[38] While certain countries use single value for intervention levels; In India, the AERB uses the concept of a range for initiating evacuation. Table, below gives the criteria used in different countries for carrying out evacuation during the off-site emergency.

Table 8.2: Off-Site Emergency Intervention Levels

Country	Intervention level [mSv]	Dose
Australia	50	Averted dose
Canada	50	Averted dose
Czech Republic	50-500	Effective dose
	100	Averted dose
Finland	50	Averted dose
Germany	100	Effective dose
Hungary	100	Effective Averted dose
Ireland	50	Averted Effective dose
Japan	50	Estimated Effective dose
	100	Estimated Equivalent Dose
India	100	Effective dose

(d) Controlling Food and Water supplies: Potentially contaminated food and water supplies are controlled, as the concentration of radionuclides in the various food items exceeds the regulatory limits prescribed by the AERB guidelines.[39]

The timely and appropriate implementation of the countermeasures during the off-site emergency can considerably reduce radiation doses to the public in the vicinity of the nuclear installation. The radiation protection countermeasures are usually implemented in the emergency planning zone (EPZ) which is spread over 16 km radius around the NPP. The radius of the EPZ can be extended depending on the radioactive release and plume dispersion patterns. The restrictions on the distribution of milk and other foodstuffs could also extend beyond the EPZ distance. For terminating the off-site emergency, the AERB has stipulated more general criteria based on the radiation effects, contamination situation, end of the release, etc. for ending the countermeasures in the affected areas.

Technical Support Systems and Emergency Response Facilities for Off-site Emergency Management

The foremost aspects of the off-site emergency planning are: a) early prediction or assessment of the extent and significance of any accidental release of radioactivity to the environment and b) rapid and continuous assessment of the accident.[40] The early detection and assessment of radioactive release during the accident can be done only by a reliable environmental radiation monitoring programme. Monitoring systems integrated with the 'real-time decision support system' can provide a coherent support at all stages of the accident (i.e. before, during and after a release has occurred), including the decision for the initiation of a wide range of countermeasures such as sheltering, evacuation, iodine prophylaxis, food restrictions, etc. (mentioned in the previous section).[41] In India, the Bhabha Atomic Research Centre (BARC) has developed: (i) Indian Environmental Radiation Monitoring System (IERMON) with data transfer facilities; (ii) GIS based Emergency Response System; and (iii) Indian Real Time Online Decision Support System (IRODOS) for detection of radioactive release, assessment of release term and further prediction of movement of radioactive plume.[42] These systems are part of the Indian nuclear emergency preparedness framework.

Under severe conditions during emergency at a NPP, the radioactive release continues for an extended period, leading to a change in the affected areas over time, and due to changing meteorological conditions (specially the wind direction). Also, in case of an evacuation, it is important that the public should avoid crossing the sectors with high radioactivity, which requires the best escape route prediction, to ensure minimum exposure. In view of these requirements, the IRODOS system provides real time online support with 72 hours meteorological and radiological forecasts, which cater to the predictive requirement for emergency planning. The IRODOS system is designed for fixed sites and takes into account the topography and local characteristics.[43]

The IRODOS system can forecast counter measures like iodine prophylaxis, sheltering, and evacuation in affected areas indicated on GIS (Geographical Information System). Figure 8.2, shows the area, villages, and cities around the NPP on GIS. The GIS provides information about the nearest public health care centres, rallying point (collection centre), sheltering locations, etc. The GIS also tracks the logistics available, like transportation, manpower, availability of iodine tablets etc. to carry out the countermeasures. The IRODOS system has been operational at the Narora Atomic Power Station (NAPS) in Uttar Pradesh for the last four years and is planned for all the NPPs in India after the Fukushima accident.[44] The NPPs in India are provided with field environmental radiation monitors (ERMs) that are placed in two rings around the NPPs. The ERMs are stand-alone, solar power-based GM detectors with GSM communication devices.

Figure 8.1: Environmental Radiation Monitoring Network

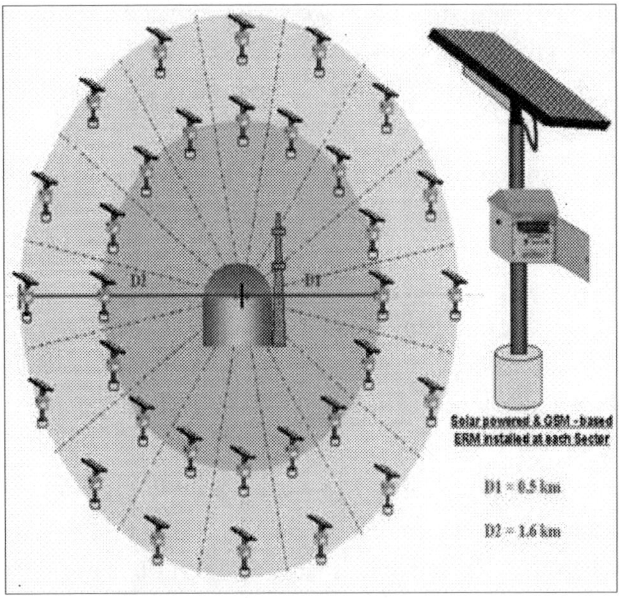

Figure 8.2: IRODOS–BARC

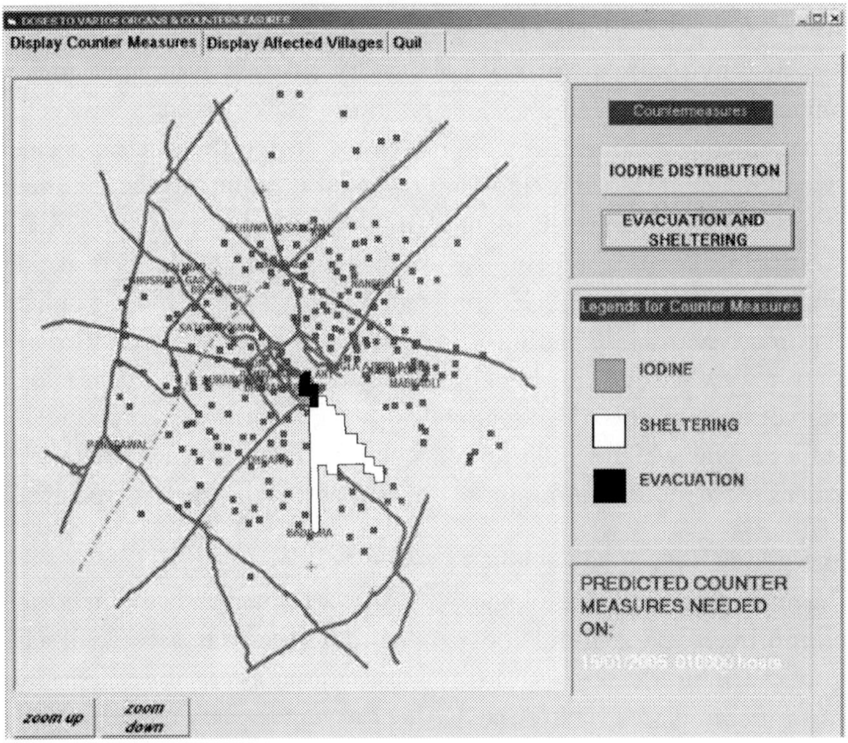

Based on the ERM readings, the IRODOS system calculates the likely source term using inverse calculations based on which the emergency measures are communicated to the first responders.[45] The BARC has also developed various methodologies and equipments for supporting off-site nuclear emergency preparedness.

1. Aerial Gamma Spectrometry System (AGSS) for aerial radiation monitoring;
2. Environmental Radiation Monitor with Navigational Aid (ERMNA)
3. Compact Aerial Radiation monitoring systems (CARMS) for remote aerial monitoring using un-manned aerial vehicles (UAV's);
4. Mobile Radiological Impact Assessment Lab (M-RIAL);[46] and
5. Portable Personal Decontamination Unit (PPDU) etc.

The Radiation Safety Systems Division of BARC has conducted several ground and aerial surveys for radiation monitoring. The aerial survey is conducted when the affected area is large and terrain is not favourable. The Aerial Gamma Spectrometry System (AGSS) is commonly used for aerial surveys. An AGSS, when fitted in an aircraft can scan a very large area and measure low levels of contamination on the ground.[47]

The Environment Survey Laboratory, at each NPP measures the contamination in food, milk, water, etc during the emergency. Based on the measurement of radioactive contamination during the nuclear accident; restrictions are imposed on food, milk and meat production in the affected areas. The Fukushima accident demonstrated the need for radiation monitoring systems capable of operating under prolonged accident conditions. The lacunas in existing technologies such as mobile radiation detection systems, GIS platforms and modelling have been identified in the safety review carried out after the Fukushima crisis. The technological shortcomings can be addressed by dedicating more resources to radiation R&D systems which would not only render the emergency preparedness systems more robust but minimise the risks during the nuclear emergencies.

Emergency Response Facilities

The mitigation of the radiological emergency immediately requires the: (i) monitoring of a large number of people from the affected areas, (ii) setting up of evacuation centres, and (iii) decontamination facilities (personnel, houses, public places, vehicles, etc.). After carrying out evacuations, it involves the isolation and confinement of contaminated areas and removal and disposal of contaminated soil. The effective response and mitigation of the radiological emergency requires resources such as:

- Trained manpower: Emergency Response Teams (ERTs).
- Equipment, radiation monitors and Protective equipment.
- Medical facilities/decontamination Centres.
- Transport and Communication facilities for the ERTs.

These facilities should be readily available at the NPPs during the emergency. To provide these resources, the DAE has set up a network of 20 ERCs all over the country including NPPs sites with a nodal ERC at the Bhabha Atomic Research Centre, Mumbai.[48] The ERCs are equipped with various types of monitoring instruments, protective gear and communication facility. It also provides for communication with agencies such as police, fire brigade, hospitals, transport services, civil defence authorities, meteorology department, NGOs and other related agencies in the area. The ERC personnel are the first responders to any major radiological emergency anywhere in the country. The ERCs at the NPPs are immediately activated by the Emergency Response Director, DAE on receiving an emergency alert.

The emergency response centre deploys various Expert Emergency Response Teams (ERTs) that are trained for specific jobs and periodically conducts mock exercises. Various types of ERTs viz. radiation monitoring and identification, aerial survey, bioassay, bio-dosimetry, medical support, etc. are being trained by the BARC for handling of nuclear or radiological emergency. The ERTs at the NPPs are equipped with transport vehicles, contamination monitors, high and low range gamma survey instruments, self reading dosimeters, TLDs, battery operated air samplers, thyroid blocking agents, first aid kits, protective overalls, portable radio communications systems, public address systems, radiation warning labels and signs, log books, identification badges, torch lights, spare batteries, field calibration of instruments, etc.[49]

The crisis following the Fukushima accident has clearly revealed the need for more trained human resources in ERTs under the existing ERCs. The NDMA is raising new battalions of the national disaster response force (NDRF). These NRDF battalions are being trained in radiological emergency response by the DAE. Emergency Response Centres currently operational, are as follows:

Table 8.3: DAE Emergency Response Centre

Sr No.	Destination	City	State
1.	Bhabha Atomic Research Centre (BARC)	Mumbai	Maharashtra
2.	Tarapur Atomic Power Station (TAPS)	Tarapur	Maharashtra
3.	Kakrapar Atomic Power Station (KAPS)	Kakrapar	Gujarat
4.	Kaiga Generating Station (KGS)	Kaiga	Karnataka
5.	Rajasthan Atomic Power Station (RAPS)	Kota	Rajasthan
6.	Atomic Minerals Directorate for Exploration and Research (AMD)	Jaipur	Rajasthan
7.	Narora Atomic Power Station (NAPS)	Narora	Uttar Pradesh
8.	Indian Rare Earths Limited (IREL)	Aluva	Kerala
9.	Uranium Corporation of India Limited (UCIL)	Jaduguda	Jharkhand
10.	Variable Energy Cyclotron Centre (VECC)	Kolkatta	West Bengal
11.	Atomic Minerals Directorate for Exploration and Research (AMD)	Shillong	Meghalaya
12.	Atomic Minerals Directorate for Exploration and Research (AMD)	Nagpur	Maharashtra
13.	Nuclear Fuel Complex (NFC)	Hyderabad	Andhra Pradesh
14.	Madras Atomic Power Station (MAPS)	Kalpakkam	Tamilnadu
15.	Atomic Minerals Directorate for Exploration and Research (AMD)	Bengaluru	Karnataka
16.	Raja Ramanna Centre for Advance Technology (RRCAT)	Indore	Madhya Pradesh
17.	Atomic Minerals Directorate for Exploration and Research (AMD)	Delhi	NCT
18.	Kudankulam Nuclear Power Plant (KKNPP)	Kudankulam	Tamilnadu
19.	Rare Materials Project (RMP)	Mysore	Karnataka
20.	Institute for Plasma Research (IPR)	Gandhinagar	Gujarat

Source: Department of Atomic Energy.

Additional ERCs are proposed at 10 locations and are likely to be operational within 1-2 years. Locations of proposed DAE-ERCs are:

1. Chatrapur, Orissa.
2. Visakhapatnam, Andhra Pradesh.

Table 8.4: Locations of Proposed NDRF ERCs

Sr No.	City	State
1.	Kolkata	West Bengal
2.	Arakkonam	Tamilnadu
3.	Pune	Maharashtra
4.	Ghaziabad	Uttar Pradesh
5.	Guwahati	Assam
6.	Mundali	Orissa
7.	Gandhinagar	Gujarat
8.	Bhatinda	Punjab

Source: Department of Atomic Energy.

Role of District Authorities and Administrative Support in Nuclear Emergency Management

To mitigate the consequences of a radiological accident, it is necessary to establish a well-coordinated and effective response at the local and national level. The National Disaster Management Authority, which was set up in 2005, has an important supportive role in handling off-site implications of accidents at the NPPs. The NDMA guidelines on 'management of radiological emergencies' have assigned roles and responsibilities to all the agencies for a series of mutually interactive, reciprocal and supplementary actions to be taken should a major accident happen.[50] Off-site emergency management is primarily the responsibility of the district and state administration hosting the NPPs in India.[51] The district commissioner is designated as the Off-Site Emergency Director (OED) and the chairs the Off-Site Emergency Committee (OEC) which is convened when the emergency is reported by the station director. The OEC comprises of the chiefs of all the concerned public service agencies, relevant to the mitigation of an emergency in the public domain.

The district commissioner declares an off-site emergency on receiving emergency alert from the NPCIL, based on which the off-site emergency plans are activated. A Crisis Management Group (CMG) for this purpose was set up in the DAE in 1987.[52] In the event of any radiological or nuclear emergency in the public domain, the CMG—which co-ordinates between the local authority in the affected area and the National Crisis Management Committee (NCMC—the executive arm of the NDMA) is activated. The CMG comprises of senior officials drawn from various units of DAE like the Nuclear Power Corporation of India Ltd (NPCIL), Bhabha Atomic

Research Centre (BARC), Heavy Water Board (HWB) and the Directorate of Purchase and Stores (DP&S).[53] It also includes a senior official from the regulatory authority, the Atomic Energy Regulatory Board (AERB). The scientific groups from BARC provide technical support to the CMG. They can provide advice and assistance in the areas of radiation measurement and protection and medical assistance to radiation affected personnel.[54] Based on the national guidelines for nuclear and radiological emergencies, the following key agencies and their responsibilities have been designated for handling nuclear emergencies:

- National Crisis Management Committee (NCMC)
- Nuclear Emergency Committee (NEC)
- National Disaster Management Authority (NDMA)
- Regional/State Emergency Response Committee
- Crisis Management Group within the DAE
- Off-site emergency director—Off-Site Emergency Committee
- Site emergency director—Site emergency committee
- Nuclear plant authority (NPCIL)
- Emergency Response Centre (ERC)
- Radiation monitoring group (BARC)
- Meteorology and environmental group
- Information and communication
- Decontamination centre
- Fire service
- Medical service
- Security (police, paramilitary unit, armed forces)
- Transport
- NGOs

For the effective implementation of off-site emergency measures, it is necessary that the whole system starts functioning once the off-site emergency is declared. It is important that the nuclear, as well as civic authorities with critical roles in executing the programme be well-prepared to perform their roles under critical conditions. Both the off-site responders and the local, district and state officials must receive intensive training on their responsibilities, and have the opportunity to practice the execution of these responsibilities in exercises that simulate real accident conditions. Between 1987 and 2010, 40 off site emergency exercises have been conducted by the respective district administrations at various locations in the country.[55] These involve direct participation by district officials like police, health, transport, etc. At the end of each of these exercises, a report is submitted to the AERB which recommends the corrective actions required to be taken by NPCIL and the district administration.

Nuclear Emergency Preparedness Post-Fukushima

The Fukushima accident caused massive socio-economic and socio-political disruptions in Japan. According to the Fukushima Reconstruction Agency, approximately 1,45,000 people were evacuated from Fukushima after the accident, of which 109,000 were from the 'Evacuation Order Area'.[56] Carrying out an evacuation on such a large-scale proved to be extremely challenging for the Japanese civic authorities in the wake of massive earthquake and tsunami, which crippled the public infrastructure across the eastern Japan. Although the triple disaster of earthquake, tsunami and nuclear meltdown is the rarest of events that could occur anywhere in the world, it nevertheless, provides important lessons for improving disaster preparedness in all the countries. In view of the situation in Japan, the Indian Government reviewed the state of nuclear emergency preparedness in the country to deal with a Fukushima like event, at the nuclear installations in India.[57] Under the directives of the country's prime minister, the National Disaster Management Authority (NDMA) carried out a structured and integrated review of emergency plans and procedures in all six states where nuclear power plants are operational.[58]

The NDMA's mandate among other things is to ensure the preparation of emergency plans at national, state, district and local level agencies. The NDMA published the first comprehensive guidelines for nuclear and radiological emergencies in 2009. In the wake of the Fukushima crisis, the NDMA conducted special mock exercises at all the NPPs in six states namely Tarapur (Maharashtra), Kalpakkam (Tamilnadu), Narora (U.P.), Kaiga (Karnataka), Kakrapar (Gujarat) and Rawatbhatta, Chittorgarh district (Rajasthan). The NDMA held a review meeting after the mock exercises to discuss the lessons learned and for preparing comprehensive off-site plans for nuclear emergencies.[59] These exercises were also aimed at understanding the actual implementation of existing NDMA guidelines into action. Based on this review, the NDMA set up two expert committees to prepare guidelines for revising the existing off-site emergency plans and to prepare a manual for conduct of mock off-site emergency exercises.[60] The NDMA, currently, faces serious challenge of notifying the guidelines, clarifying the responsibilities of local agencies and also garnering support of states hosting NPPs.[61] The review of the off-site emergency preparedness programme revealed important lacunas in terms of lack of public infrastructure in areas around NPPs and the training of disaster relief personnel.

Infrastructure Improvements

The NDMA has accorded the highest priority to developing infrastructure in the form of equipment, facilities, services, etc. that are necessary to carry out emergency measures in the event of a severe nuclear accident. While most of the technical infrastructure has been developed by the Bhabha Atomic Research Centre (BARC), the infrastructure for public use remains a neglected area in off-site emergency preparedness and planning in India. Connectivity by roads, footbridges in remote villages and narrow pavements has been identified as a serious problem during the off-site emergency exercises in almost all the NPP sites. Similarly, basic infrastructure like sanitation facilities, electricity supply, drinking water in school and government buildings and health centres also needs to be improved in the emergency zones. The review suggested improving infrastructure facilities like roads, bridges, schools, hospitals, etc. in areas nearby NPPs and the ensuring of regular power supply to these areas.

The NDMA has approached the state governments to prepare plans for developing local infrastructure in the areas around NPPs. The states have also been asked to set up development plans for covering the 16 km emergency zone in the vicinity of the NPPs, which have not been implemented in the past. Two years after Fukushima accident, the Karnataka state government has prepared a development plan costing Rs 200 crore for upgrading the public infrastructure in the emergency planning zone around Kaiga NPP.[62] The five other states hosting NPPs, have been asked to implement similar development plans in areas around the NPP sites. The NDMA also suggested that the emergency preparedness exercises should be held at least once in a year instead of once in two years.

Nuclear Emergency Health-Care

While reviewing the off-site emergency exercise, it was observed that the existing health facilities at the hospitals of central and state governments in the districts hosting NPPs need thorough strengthening. It also highlighted the importance to identify and train human resources specialised in handling radiation injuries as part of nuclear emergency management, in secondary and tertiary healthcare facilities in the prioritised districts. For handling off-site emergencies, quick response medical teams must be pre-positioned in all districts having NPPs and those likely to be affected. It was recommended that, at least one hospital in each district should be identified, equipped and upgraded to handle excessive radiation exposure and injuries to the people. To address these issues, the Ministry of Health and Family Welfare (MoH&FW) was asked to constitute a Task Force which developed a road map for enhanced medical preparedness in the country to deal with nuclear and radiological emergencies.

Table 8.5: Nuclear Emergency Response Facilities[64]

Sl. No.	Facilities	Existing	Proposed	Projected Requirement
1.	ERCs	18+2(BARC)	10 NDRF* BNs	100
2.	HAZMAT Vehicles	9*	6	35
3.	CBRN Surveillance Vehicles	1	9	35
4.	Mobile Decontamination Facility	Nil	10	35

The Task Force assessed the capacities in the country which include; human resources to handle radiation injuries, pre-positioning, quick response by the medical team in areas likely to be exposed, existing health facilities in districts hosting NPPs, strengthening of central government and state government hospitals in six metros and in the vicinities of other vulnerable districts, and also the need to upgrade and equip at least one hospital in every NPP hosting district to deal with radiation emergencies . The task force recommended setting up of a centre of excellence at Safdarjung hospital, New Delhi, besides various centres for tertiary level medical management and research on nuclear emergencies in different parts of the country including NPP sites.[63] Among other things, the task force recommended stock piling of drugs and equipment in designated hospitals, generating awareness among medical professionals and general public regarding nuclear emergencies, providing post–emergency psycho-social care and strengthening ongoing research programmes.[65]

More Realistic Drills and Training

Effective implementation of the off-site emergency plans requires the concerned agencies to be well-prepared to perform their roles under extreme conditions. On-site personnel and off-site responders along with officials from local, state, and central agencies must receive in-depth training about their responsibilities, and participate in exercises that simulate real accident conditions. The DAE has initiated an ambitious programme of training, drills, and exercises to ensure that the on-site and off-site agencies are well-prepared for handling nuclear emergencies. The DAEs's off-site emergency exercise programme has been suitably modified to include the lessons from Fukushima; to handle contamination of large areas due to large-scale radioactive fallout. The safety reviews undertaken at the NPPs also addressed problems such as failures at multiple units, and emphasized the need to prepare for a protracted period in case, of station black-out (SBO) scenarios.

The DAE is setting up more emergency response centres to provide additional support in terms of equipment and human resources. Given the geographical expanse of the country, only 30,000 people are trained in disaster management, and this number needs to increase. To address this challenge, the NDMA has initiated a de-centralised programme to train more personnel in emergency rescue, relief and in radiological decontamination procedures. In addition, the NPCIL has also launched extensive public awareness programmes in villages and townships in the emergency planning zones, to inform people about the protective measures during the emergency at NPPs.

Conclusion

In the aftermath of the Fukushima Dai-ichi accident, there is a significant need for improving the public understanding about radiation-risks with regard to off-site emergency preparedness in India. With adequate emergency preparedness measures in place, the hazards of radiation exposure can be mitigated. The AERB has developed a robust regulatory framework to declare and handle off-site emergencies at the NPPs. The nuclear authorities have put in place a diverse technological system to detect and handle radiological emergencies. These technical systems facilitate critical decision making for implementing the radiation protection counter-measures and mitigating the accident consequence. The off-site emergency plans and procedures of the state and district administration are being revised to include the lessons learned from the special off-site emergency exercises held after Fukushima crisis. Reviews of the off-site emergency mitigation exercises revealed that the capabilities of the district and state administration for handling off-site emergencies need to be strengthened, in terms of necessary infrastructure and trained resource personnel. The challenges involved in managing and mitigating the off-site emergencies can thus, be addressed by sustained efforts to enhance and upgrade the resources and infrastructure for nuclear emergencies.

NOTES & REFERENCES

1. The Fukushima accident caused massive socio-economic and socio-political disruption in Japan. After the Fukushima accident, there has been widespread speculation about the state of nuclear emergency preparedness in India. Dr A. Gopalakrishnan, former Chairman, AERB, criticised India's readiness to deal with emergencies as merely a paper exercise. Civil society groups have also referred to the comptroller and auditor general's report on performance and audit of the activities of Atomic Energy Regulatory Board (AERB) that was released on August 23, 2012 which stated that, "emergency exercises carried out at various sites highlighted inadequate emergency preparedness."
2. The AERB safety guides discuss in detail various types of radiological emergencies. This papers as mentioned in the introduction is limited to the discussion on off-site emergency preparedness in India.
3. Zeigler, Donald J. and Johnson Jr., James H. (1989), "Information Planning and the Geographical Context of Radiological Emergency Management", *Transactions of the Institute of British Geographers*, New Series, Vol. 14, No. 3 (1989), pp. 350-363.
4. Ibid.
5. Brunn, S.D. and Johnson, J.H., Jr. and Zeigler, D.J. (1979), "*Final report on a social survey of Three Mile Island area residents*", Michigan State University: East Lansing, MI.
6. Ibid, Zeigler, Donald J. and Johnson, Jr., James H. (1989).
7. Nuclear Regulatory Commission, "History of Emergency Preparedness Response in USA", URL: http://www.nrc.gov/about-nrc/emerg-preparedness/history.html, accessed on April 20, 2013.
8. US-NRC, "Planning Basis for the Development of State and Local Government Radiological Emergency Response Plans In Support of Light Water Nuclear Power Plants", *NUREG-0396, EPA 520/1-78-016*, pp. 01-25, URL: http://pbadupws.nrc.gov/docs/ML0513/ML051390356.pdf, accessed on May 10, 2013.
9. Ibid, US-NRC, *NUREG-0396, EPA 520/1-78-016*, p.25, accessed on May 10, 2013.
10. Ibid, US-NRC, *NUREG-0396, EPA 520/1-78-016*, pp. 1-37, accessed on May 10, 2013.
11. Nuclear Regulatory Commission, "NRC History of Emergency Preparedness & Response", URL: http://www.nrc.gov/about-nrc/emerg-preparedness/history.html, accessed on April 25, 2013.
12. Demin, V.F. and Yatsalo, B.I. (1998), "Chernobyl" Lessons Learned for Post-Emergency Response", *IRPA International Convention*, Paper No. P-11-231.
13. See, Atomic Energy Regulatory Board, "Preparation of Off-site Emergency Plans for Nuclear Facilities", *Safety Guide No. AERB/SG/EP-2 and Safety Guide No. AERB/SG/O-6.*
14. Ibid, See, AERB/SG/EP-2: Preparation of Off-site Emergency Plans for Nuclear Facilities and AERB Safety Guide No. AERB/SG/O-6.
15. Atomic Energy Regulatory Board (2000), "Regulation of Nuclear and Radiation Facilities" *Safety Code AERB/SC/G, 2000.*
16. Atomic Energy Regulatory Board (2008), "The Safety Code "Safety in Nuclear Power Plant Operation" *Safety Code AERB/SC/O, 2008.*
17. Atomic Energy Regulatory Board (2000), "Role of the Regulatory Body with Respect to Emergency Response and Preparedness at Nuclear and Radiation Facilities" *Safety Guide AERB/SG/G-5, 2000.*

18. Atomic Energy Regulatory Board (2000), "Preparedness of the Plant Management for Handling Emergencies at NPPs" *Safety Guide AERB/SG/O-6, 2000.*
19. Atomic Energy Regulatory Board (2000), "Preparedness of the Plant Management for Handling Emergencies at NPPs" *Safety Guide AERB/SG/O-6, 2000.*
20. Atomic Energy Regulatory Board (1999), "Preparation of Site Emergency Plans for Nuclear Installation" *Safety Guide AERB/SG/EP-1,* 1999, and Atomic Energy Regulatory Board (1999), "Preparation of Site Emergency Plans for Nuclear Installation" *Safety Guide AERB/SG/EP-1, 1999.*
21. Atomic Energy Regulatory Board (2000), "The safety manual on Radiation Protection in Nuclear Facilities" *AERB/NF/SM/O-2, Rev. 4, 2005.*
22. Government of India, Department of Home Affairs, "The Disaster Management Act, 2005" pp. 1-2.
23. National Disaster Management Authority (2009), "Management of Nuclear and Radiological Emergencies Guidelines", February 2009. Pp. 1-132
24. Ibid, p.18.
25. Nuclear emergencies at the NPPS are rated on the International Nuclear and Radiological Event Scale (INES), from Level-1 (anomaly) to Level–7 (major accident) depending on the radiological release and its impact. The Chernobyl and Fukushima accident were rated at Level-7. In India, out of 156 events reported from various NPPs during year 2007-11; as many as 140 events belonged to Level-0 (i.e. no safety significance) and the remaining 16 were of Level-1. In the past, there were two major events of safety significance: (i) the fire incident in turbine building at Narora Atomic Power Station (1993, INES Scale Level-3) and (ii) unintended power excursion in Kakrapar Atomic Power Station (2004, INES 11 Scale Level-2. There were no radiological consequences for the workers, public or the environment from these events
26. India signed the Convention on Nuclear Safety on September 1994 and subsequently ratified it on June 29, 2005. Being signatory to the convention, India has submitted 'Country Report' respectively in year 2007, 2009, and 2012 which describes India's compliance with the provisions of CNS. These reports can be accessed on CNS's web portal.
27. See, Safety Guide AERB/SG/EP-2, 1999.
28. See, Safety Guide AERB/SG/EP-2, p.20.
29. Ibid.
30. Pushparaja, Pradeepkumar, K.S., Sharma, D.N. and Kushwaha, H.S. (2006), "Philosophy and principles of Radiation Protection, Security of Radioactive Sources and Preparedness for Response to Radiological Emergencies", *BARC Newsletter,* Issue no. 275, December 2006
31. Abani, M.C (2011), "Nuclear and Radiological Emergencies: Preparedness and Response", *Defence and Security Alert,* June 2011, pp. 8-9.
32. Ibid, p.28.
33. International Atomic Energy Agency, "Arrangements for Preparedness for a Nuclear or Radiological Emergency", *Safety Guide, No. GS-G-2.1,* pp. 5-7.
34. Ibid, p. 22.
35. AERB Safety Guide AERB/SG/EP-2, Annexure Table A-5, p.25.
36. AERB Safety Guide AERB/SG/EP-2, Annexure Table A-4, p.24.
37. AERB Safety Guide AERB/SG/EP-2, Annexure Table A-7, p.27.
38. OECD (2013), "Short-term Countermeasures in Case of a Nuclear or Radiological Emergency", ISBN 92-64-02140-X.
39. Ibid, AERB Safety Guide AERB/SG/EP-2, Annexure Table A-1 & A-2, p.27.
40. A. Vinod Kumar, R.B. Oza, et.al (2008), "Indian Real Time Online Decision Support System (IRODOS)", *BARC Newsletter,* Issue No. 292, May 2008.
41. W. Raskob, "The RODOS System: Decision Support For Nuclear Off-Site Emergency

Management In Europe", *IRPA, Paper No. T-16-3, P-11-292.*
42. Ibid, *R.B. Oza, R.B. Oza, et al.*
43. Ibid, *R.B. Oza, R.B. Oza, et al.*
44. AERB One-Day Meeting, "Review of Off-site Emergency Preparedness and Response Plan of Indian NPPs based on Experience of Fukushima Nuclear Accident," *AERB Newsletter*, 24 (2), July-December, 2011, p. 13.
45. Pushparaja, K.S. Pradeepkumar, D.N. Sharma and H.S. Kushwaha, "Philosophy and Principles of Radiation Protection, Security Of Radioactive Sources and Preparedness for Response to Radiological Emergencies", *BARC Newsletter,*
46. K.S. Pradeep Kumar (2013), '*National level preparedness for nuclear and radiological emergencies",* BARC, Mumbai.
47. Ibid.
48. Lok Sabha unstarred question no. 1827, to be answered on 10.08.2011, radiation emergency response centres, URL: http://www.dae.gov.in/parlqa/2011/monsoon2011/1827_lsus100811.pdf
49. Lok Sabha – Unstarred Question No-1827, Department of Atomic Energy, URL: https://dae.nic.in/writereaddata/1827_lsus100811.pdf.
50. Ibid.
51. Ibid, p.xxviii.
52. Department of Atomic Energy, *"Nuclear Emergency Response"*, http://dae.nic.in/?q=node/37, accessed on May 20, 2013.
53. Ibid.
54. Department of Atomic Energy, "Emergency Response Plans in DAE: An Overview", URL: http://www.dae.gov.in/ni/ninov02/ecr.htm, accessed on May 18, 2013
55. Ibid, Emergency Response Plans in DAE: An Overview.
56. Fukushima Reconstruction Agency, "*The Status in Fukushima: Progress to Date*", URL: http://www.reconstruction.go.jp/english/topics/progress-to-date.html, accessed on May 25, 2013.
57. PM reviews atomic disaster preparedness, *PTI*, June 01, 2011, URL: http://www.thehindubusinessline.com/economy/pm-reviews-atomic-disaster-preparedness/article2067954.ece, accessed on May 22, 2013.
58. Press Release, "Development of Infrastructure for Rehabilitation of Families In and Around Kaiga Nuclear Plant", *National Disaster Management Authority*, URL: http://ndma.gov.in/ndma/pressrelease/PressreleaseKaiga25Jan2012.pdf, accessed on May 22, 2013.
59. Ibid.
60. National Disaster Management Authority, Office Order No: NDMA/NDRF/TNR/402-Pt.1/2011/3697 dated August 26, 2011. The author accessed this document from the office of Dr. B Bhattacharjee, Member, NDMA on September 05, 2012.
61. Shashidhar Reddy, "*Nuclear Disaster Management in India*", talk delivered at the Institute for Defence Studies and Analyses (IDSA) on January 20, 2012. Mr Shashidhar Reddy, Vice Chairman National Disaster Management Authority (NDMA) during his public addresses stated that, "there is a need to dwell more on failures and consequences of inaction in order to secure a safe nuclear future for India".
62. Press Release, *Development of Infrastructure for Rehabilitation of Families In and Around Kaiga Nuclear Plant,* National Disaster Management Authority.
63. Press Briefing by Shri M. Shashidhar Reddy, "Special Meeting of NDMA on India's Disaster Preparedness in the wake of Recent Disasters in Japan", June 01, 2011, NDMA, URL: http://ndma.gov.in/ndma/pressrelease/pmmeeting.pdf.
64. Shashidhar Reddy, "*Nuclear Disaster Management in India*", from the unpublished presentation made by author at the Institute for Defence Studies and Analyses on January 20, 2012.

CHAPTER 9

Civil Nuclear Liability in India: Challenges and Prospects

G. Balachandran

Introduction

With the increasing emphasis on the environmental aspects of energy generation and use, especially with regard to controlling and decreasing carbon emissions, nuclear energy is once again attracting global attention as an efficient alternative to other forms of electricity generation. However, the efficiency and attractiveness of nuclear energy is being rigorously contested and contrasted against the potential hazards that it signifies. In fact, the tirade against nuclear energy is gaining momentum, particularly after the Fukushima incident in 2011.

Against this backdrop, India has ambitious plans to increase the share of nuclear electricity generation to the levels prevailing in some developed countries i.e. 40 per cent and above, in view of the enormous demand for power that will be required to drive India's economic growth in the coming decades. These plans have received further impetus with the amendment to NSG guidelines that have made civil nuclear commerce between India and NSG members possible after a gap of nearly 20 years. It is expected that the number of civil nuclear reactors operating in India will increase by leaps and bounds in the coming years, especially as the Indian experience of operating of nuclear power plants (NPP) has been quite successful and incident free in the past. This has been partly due to the technology used as well as the safety measures practised exercised over NPP operations, and well trained operating personnel.

Nevertheless, as the number and types of NPPs increase, India needs to modify its domestic legal mechanism to cater to any eventuality leading to major accident at nuclear power plants that involves settling the large number of compensation claims arising from such an eventuality. While, it is necessary that such legal mechanism should have the interests of the public as its primary focus; it should also at the same time facilitate nuclear commerce and not obstruct it. In the light of the above, this chapter outlines basic principles underlining the international civil nuclear liability framework, with specific focus on issues such as suppliers' liability and right to recourse. It then examines the prospects and challenges presented by the India's Civil Liability for Nuclear Damage Act of 2010.

The International Nuclear Liability Regimes

There are currently 30 countries that operate 435 nuclear power plants (NPPs.[1] Of these, 29 countries have 434 NPPs in operation, and also have some sort of nuclear liability act in force in their territory. This is so either because of adherence to some international liability regime—the IAEA's Vienna Convention for Nuclear Damage of 1963 or the OECD's Paris Convention on Third Party Nuclear Liability in the field of Nuclear Energy of 1960—or as a result of enacting a national liability law. Twenty-two of the 29 countries are party to one or the other of the two international conventions. The other seven—Canada, China, India, Japan, Republic of Korea, South Africa and Switzerland have national laws on nuclear liability. The US is slightly different, wherein although the supplier is legally liable; he is not required in any manner to compensate the operator. Only Pakistan, that is operating two NPPs, is not a member of either international convention, nor does it have any national legislation.

The enormous power of nuclear technology was first witnessed in Hiroshima and Nagasaki. Therefore when the first peaceful applications started to fructify in form of NPPs in the 1950s, policy makers felt that the issue of civil liability for damage caused by incidents in nuclear installations called for special statutory provisions, that differed from the specific rules that had already been formulated to govern third party liability for damage caused by incidents causing hazards to the population and environment in general. Given the special hazards of nuclear accidents and the possibility that such accidents may involve the nationals of more than one country, it was felt that an international regime for nuclear liability would be a welcome step.

Paris and Vienna Conventions

The need for international regulation was first acknowledged by the states engaged in the development of nuclear energy. These included the members of the then Organisation for European Economic Co-operation (OEEC), which was later reconstituted as the Organisation for Economic Co-operation and Development (OECD), and the European Atomic Energy Community (Euratom). As a result, the 1960 Paris Convention on Third Party Liability in the Field of Nuclear Energy entered into force on April 1, 1968. This Convention was followed by the 1963 Brussels Convention, supplementary to the Paris Convention, which entered into force on December 4, 1974. The need for a uniform nuclear liability regime at the global level was also felt and, on May 21, 1963, the Vienna Convention on Civil Liability for Nuclear Damage was adopted under the auspices of the International Atomic Energy Agency (IAEA). The 1963 Vienna Convention entered into force on November 12, 1977.[2]

The existing international nuclear liability regime is thus based on the Convention on Third Party Liability in the Field of Nuclear Energy of July 29, 1960, as amended by the Additional Protocol of January 28, 1964 and by the Protocol of November 16, 1982 (Paris Convention) and the 1963 Vienna Convention on Civil Liability for Nuclear Damage (Vienna Convention).[3] The modern IAEA nuclear liability regime essentially incorporates the principles embedded in these agreements, which include:

(a) The operator of a nuclear installation is exclusively liable for nuclear damage. All liability is channelled on to one entity, namely the operator of the nuclear installation where the nuclear incident occurs. Under the various conventions, the operator — and only the operator — is liable for nuclear incidents, to the exclusion of any other person or entity. Two factors have motivated this exclusive liability of the operator, as distinct from the position as per the ordinary law of torts. Firstly, because it is desirable to avoid difficult and complicated legal cross-actions to establish in individual cases who is legally liable. Secondly, such exclusive liability obviates the necessity for all those who might be associated with the construction or operation of a nuclear installation other than the operator himself, to also take out insurance, and thus allows a concentration of the insurance capacity available.

(b) Strict (no-fault) liability is imposed on the operator.

Notwithstanding the relative safety of NPPs, nuclear accidents are marked by certain unique features. First of all, the damage caused by ionising radiation to living cells, especially human cells, may not be immediately recognisable; it may remain latent for a long time. Secondly, the detrimental effects of a major nuclear accident may extend far beyond the territory of a state as was seen in case of Chernobyl accident. Third, given the nature of NPP accidents, the proof of causation depends upon sophisticated scientific evidence given the nature of NPP operations. Fourthly and finally, in case of offsite nuclear damage, it is a unilateral accident i.e. one which the plaintiffs, will neither be able to, nor can in any manner prevents as the defendant or the operator of the NPP has the sole capacity to control the risk of accident. For all these reasons a nuclear liability bill should incorporate the principle of strict liability. The operator of a nuclear installation should be held liable, regardless of fault. The plaintiff need not have to prove negligence or any other type of fault on the part of the operator.

(c) Exclusive jurisdiction is granted to the courts of one country, to the exclusion of the courts in other countries. The general rule is that a court of the contracting party in whose territory the nuclear incident occurs has jurisdiction. If suits arising out of the same incident were to be tried and judgments rendered in the courts of several different countries, the problem of assuring equitable distribution of compensation might become insoluble. Within the country, one single competent forum should initiate all actions—including direct actions against insurers or other guarantors and actions to establish rights to claim compensation—against the operator of the same nuclear incident.

(d) Liability is limited in amount and in time. In the absence of a limitation of liability, the risks could, in the worst case scenario involve financial liabilities greater than any hitherto encountered and it would be very difficult for operators to find the necessary insurance or financial security to meet the risks. As to the limitation of time, bodily injury caused by radioactive contamination may not become manifest for some time after the exposure to radiation has actually occurred. The legal period during which an action may be brought is therefore a matter of great importance. Operators and their insurers or financial guarantors will naturally be concerned if they have to maintain, over long periods of time, reserves against outstanding or expired policies for possibly large but unascertainable amounts of liability. On the other hand, it is unreasonable for victims whose damage manifests itself late to find no provision has been made to compensate them. A further complication is the difficulty of proof involved in establishing or denying that delayed damage was, in fact, caused by the nuclear incident. A compromise has necessarily been arrived at between the interests of those suffering damage and the interests of operators.

Convention on Supplementary Compensation (CSC)

In general these two conventions either placed a limit on the operator's liability or a limit on the liability cover to be taken by the operator by way of insurance or other financial security. While these principles may be justified because of the need to encourage nuclear activities, however, in case of an accident, the liability cap may not be enough to for pay the full damage. Most countries have therefore sought to overcome the damage occurring over and beyond the liability cap by making provisions to cover such damage from public funds.

The Brussels Convention, which is only for states party to the Paris Convention, established a regional system of additional coverage of damage provided partly by the Installation State and partly by all the Contracting Parties together to provide supplementary compensation in case of nuclear accidents.[4] Unlike the Paris Convention, the 1963 Vienna Convention does not expressly envisage the possibility for the Contracting Parties to establish a system of additional compensation for damage exceeding the operator's liability.[5]

After the Chernobyl incident, the idea of supplementary funding at the global level attracted renewed interest. Work centred at first on the expansion of a world convention which would supplement both the Vienna Convention and the Paris Convention, and would only be open to states party to either one of those Conventions. However, after much discussion it was eventually decided to expand the scope of the convention to include not only the states party to the Paris Convention or the Vienna Convention but also the states party to neither convention, provided their legislation conformed to certain basic principles of nuclear liability. This resulted in the Convention on

Supplementary Compensation (CSC). The Preamble to the 1997 Convention on Supplementary Compensation makes it clear that the purpose of the new Convention was the establishment of a worldwide liability regime "to supplement and enhance" measures provided not only in the Vienna Convention and in the Paris Convention, but also in national legislation "consistent with the principles of these Conventions". For this reason, the Convention sets out specific provisions on civil liability for nuclear damage in an Annexure thereto, and Article II.3 states that the Annexure constitutes an integral part of the Convention.[6]

The Annexure provides, in particular, for: "absolute" and exclusive liability of the operator of a nuclear installation (Article 3); limitation of liability in amount and/or of liability cover by insurance or other financial security (Articles 4 and 5); limitation of liability in time (Article 9), matters already elaborated in the earlier section.[7]

Articles XVIII and XIX make it clear that instruments of ratification, acceptance or approval, as well as instruments of accession, shall be accepted only on the part of states party to either the Vienna Convention or the Paris Conventions or of states declaring that their national law complies with the provisions of the Annexure.[8]

The Indian Civil Liability for Nuclear Damage Act, 2010

India was one the countries that very early in the nuclear age realised the potential of the peaceful applications of nuclear energy and embarked on a well conceived plan to reap the benefits of this new science in the spheres of agriculture, medicine, and energy. It was, therefore, not surprising that India was one the first countries to build and operate nuclear power plants (NPPs). However, the 1962 Atomic Energy Act under which the Indian nuclear programme has developed is silent on the subject of compensation to victims of a nuclear accident or incident.[9] Till the passage of the Civil Liability for Nuclear Damage Bill in 2010, there was no regulatory mechanism or legislative provision in India for dealing with claims arising out of a nuclear incident.[10]

Nevertheless, the Department of Atomic Energy began the exercise for crafting such a liability legislation in 2000 when the Board of Research in Nuclear Studies commissioned Prof. V.B. Coutinho of Government Law College, Bangalore and Prof. S. Rajagopal of National Institute of Advanced Studies, Bangalore to undertake a study on the nuclear liability regime and related issues.[11] The Report which was submitted in November 2001 examined the international legal regimes in force and suggested that it was time for India to establish a domestic legal mechanism to provide compensation to victims of a nuclear incident. It was also felt that there could be trans-boundary effects of a nuclear accident, an aspect which is relevant in the case of the Kudankulum Nuclear Power Station.

However, the government did not proceed with such legislation till 2010, when it became evident that with the relaxation of the NSG guidelines for nuclear commerce with India, foreign nuclear suppliers were keen to enter India for providing nuclear power reactors and nuclear fuel and enable the Indian nuclear power programme to expand. The government felt that it was essential to provide nuclear liability protection to enable the nuclear industry, both domestic and foreign, to participate in the planned expansion. The Legislation was also a necessity to become a member of an appropriate international liability regime, in this case the CSC. Accordingly, the bill that was introduced in the Parliament fully complied with the requirements of the CSC and in particular the provisions of its Annexure and particularly Article 10 of the Annexure on the Right of Recourse, which stated clearly that, "National law may provide that the operator shall have a right of recourse only: (a) if this is expressly provided for by a contract in writing; or (b) if the nuclear incident results from an act or omission done with intent to cause damage, against the individual who has acted or omitted to act with such intent."[12]

Right of Recourse in this case means that the operator will be liable to first, compensate the victims in case of a nuclear accident, and after having compensated the victims in line with its own liability, will have the right to sue the suppliers if there is enough proof of faulty/sub-standard supplies being the primary cause of the accident. The Indian draft bill had an additional circumstance under which the Right of Recourse could be exercised. This was: "where the nuclear incident has resulted from the wilful act or gross negligence on the part of the supplier of the material, equipment or services, or of his employee" (Sec. 17(b) of the Indian draft bill). Prima Facie it would appear that Sec. 17(b) of the Indian nuclear liability bill goes beyond the requirements of all the three conventions—Paris, Vienna and CSC and is hence in violation of the three conventions.

However, an examination of the nuclear liability laws in countries that are not party to either of the international conventions shows that the Indian draft formulation of the operator's right of recourse was not unusual or unique—South Korea, for example, has an almost similar condition. Hence while Section 17(b) of the Indian draft bill may be held contrary to the requirements of CSC and the international conventions, that by itself is no bar against foreign suppliers supplying nuclear items to India. South Korea, for example, has vendors from Canada, France, and USA etc supplying nuclear items without being unduly perturbed by the Operator's right of recourse against the supplier, unless, of course, the suppliers have special agreements regarding rights of recourse.

Liability Limits

The maximum liability in case of the current Nuclear Liability Act is SDR 300 million in rupees and the operator liability is different for each nuclear incident. According to Section 6(2) of the Nuclear Liability Act, the liability of an operator in respect of nuclear reactors having thermal power equal or above 10 MW shall be Rs 1500 crores, whereas that of spent fuel reprocessing plants is Rs 300 crores.[13] Moreover, for research reactors with thermal power below 10 MW, or those that have fuel cycle facilities other than spent fuel reprocessing, and also nuclear material transportation, the liability limit for an operator is Rs 100 crore. The limits under the current act, fall somewhere in the mid range of the global pattern. More specifically, of the 28 other countries that operate NPPs and have some liability laws; only in four countries is the maximum liability for a nuclear accident higher than the Indian limit of SDR 300 million. In 24 countries the maximum liability is less than or equal to the Indian maximum liability limit. As for the operator liability, the Indian limit is half that of the others. Fourteen of the other 28 countries have a higher limit for the operator's liability. The Table below shows the liability limits under the conventions currently in force.

Table 9.1: International Conventions on Civil Nuclear Liability

Convention	Contributor	Amount
Paris	Nuclear Operator	SDR 5 Million
BSC	Nuclear Operator or Installation State Collective State Fund	SDR 175 Million SDR 125 Million
Total NEA regime		SDR 300 Million
Vienna Convention	Nuclear Operator	Not less than US$ 5 Million

However, although Sec. 6(1) of the liability act, states that "The maximum amount of liability in respect of each nuclear incident shall be the rupee equivalent of three hundred million Special Drawing Rights", there is no further mention of this amount or reference to Sec. 6(1) anywhere else in the act. On the contrary, Sec. 7(a) is very clear that:

> The central government shall be liable for the nuclear damage in respect of a nuclear incident where the liability exceeds the amount of liability of an operator under sub-section (2) of section 6, to the extent such liability exceeds such liability of the operator.
>
> This implicit understanding of unlimited maximum liability in case of a nuclear incident is reinforced by the fact that the bill has no provision for dealing with a situation wherein the total liability exceeds the maximum liability referenced in sub-section 6(1).

In addition to the Sections 6 and 7 of the Act which appear somewhat contradictory and ambiguous, another major caveat in the Nuclear Liability Act is given in Section 17 which deals with the previously mentioned operator's Right of Recourse. This is discussed in the following sub-section.

Section 17(b) Implications: Foreign and Domestic

For a variety of reasons, mostly politico-ideological, Sec. 17 (b) of the draft bill when passed by the Parliament was amended to read:

> "(b) the nuclear incident has resulted as a consequence of an act of supplier or his employee, which includes supply of equipment or material with patent or latent defects or sub-standard services;" omitting the qualifier "the wilful act or gross negligence."This has some serious implications for the future development of India's nuclear power programme.

(i) Foreign

In the period immediately following the NSG exemption that was granted to India in September 2008, India entered into a number of bilateral cooperation agreements—USA, France, Canada, South Korea, Kazakhstan etc.—with the expectation of jump starting the Indian nuclear power programme with the import of a number of light water reactors. Various sites were allotted to USA, Russia and France for 26 light water reactors with a capacity of 30,000 MWe and work was expected to start on eight plants with a capacity of 10,500 MWe during the XII Plan period (2012-17).[14] Yet even now—after a gap of more than four years since the NSG exemption was granted not a single commercial contract has been signed for the import of any light water reactor—even for the two additional Russian reactors at Kudankulum where already two such reactors are ready to be commissioned. The primary reason for this has been Sec. 17(b) of the Indian Act. All foreign vendors have expressed their unwillingness to supply reactors to India because of this provision—some have expressed their reservations openly e.g. the US; while others, especially the French, have been making their reservations known privately to the Indian government and the Indian nuclear power operator. A case in point is the India-France agreement for nuclear cooperation signed in 2008. This agreement states that:

> The Parties agree that, for the purpose of compensating for damage caused by a nuclear incident involving material, nuclear material, equipment, facilities and technology referred to in Article IX, each party shall create a nuclear liability regime based upon established international principles. (Art. VIII (2) of the India-France Cooperation Agreement on the development of peaceful uses of Nuclear Energy).[15]

The French had for long insisted on such a liability regime before agreeing to supply nuclear items. In fact, China put in place such a liability regime in March 1986, only to address the concerns of foreign suppliers, mainly the French, who were to work on the construction of the second Chinese NPP at Daya Bay.[16] Hence notwithstanding the contentions of the critics about the French being willing to supply equipment without a nuclear liability bill, so far no responsible French official or nuclear supplier, such as AREVA for example, has ever made such a claim. This is the main reason why in spite of an India-France agreement on nuclear cooperation, the contract on the Jaitapur NPP in India has not yet been concluded.

What about Russia and Kudankulam III and IV? This is an interesting case because of the history of Russia's engagement in nuclear commerce with France and Germany. In June 2000, the governments of the Russian federation and the French Republic signed a bilateral agreement to "govern issues of liability for nuclear damage in the event of a nuclear incident within the territory of the Russian Federation that results from deliveries from the French Republic to nuclear installations in the Russian Federation". Specifically Article III of the Agreement stated:

1. The Russian Party shall bring no claims against the French Party or against suppliers on grounds of nuclear damage resulting from a nuclear incident which has taken place within the territory of the Russian Federation.[17]

2. The Russian Party shall grant the French Party and the suppliers' appropriate legal protection and shall exempt them from liability for damages in the event of claims by third parties on grounds of nuclear damage resulting from a nuclear incident which has taken place within the territory of the Russian Federation.[18]

Earlier Russia had signed a similar agreement with Germany, in June 1998, which set out the rules of liability applicable to a nuclear incident in the Russian Federation involving German equipment supplies. In such a case, the Russian Federation agreed not to institute liability proceedings against Germany or against any German supplier, and to ensure that they would receive sufficient legal protection and not be held liable in respect of claims made by third parties. This agreement, by mutual consent, lapsed when the Russian federation ratified the Vienna Convention.

It is unlikely that the Russian supplier would agree to nuclear damage liability damage in India in the absence of such a specific condition being part of the India-Russia agreement on Kudankulum. It would be interesting to know if the Kudankulum contract specifically addresses the issue of supplier liability. If it does go beyond the standard warranty clauses, then a claim against the Russian supplier would prove infructuous. Thus it is most likely that even under current conditions the Indian public will have no recourse against a foreign supplier. Therefore, as it now seems apparent, the Russians have the same problem with the Right of Recourse as stated in the Nuclear Liability Act. So far, once again, in spite of many intergovernmental agreements not a single commercial contract has been signed by India and Russia on Kudankulam III and IV.

The Russian objections to supplying nuclear reactors for Kudankulum III and IV are based on entirely different grounds. Atomstroyexport (ASE), the authorised organisation of Russian Federation, in their commercial offer to Nuclear Power Corporation of India Limited (NPCIL) in respect of Kudankulam Units III and IV had stated that NPCIL being the operator is fully responsible for any damage caused to any person and property as a result of a nuclear incident occurring at the NPP. The ASE had mentioned that their plea was based on the principle of continuity of cooperation in the construction of the KKNPP as per the agreement signed between the USSR and Republic of India on November 20, 1988, supplemented on June 21, 1998 and in line with the provisions of Intergovernmental Agreement dated December 5, 2008.

The Attorney General of India had then opined that the decision to not exercise the right of recourse under Section 17 of Civil Liability for Nuclear Damage Act, 2010 at the time of the signing of the Techno-Commercial Offer (TCO) would be legally permissible.[19] However, such a decision by NPCIL would have long term political and economic implications in the future. Politically, it would be difficult for NPCIL to exercise such a choice since it is a PSU (Public Sector Unit) under the department of energy and cannot be seen to invalidate an act passed by the parliament. Further such a move will have other suppliers seeking similar concessions. Hence, at this stage it seems unlikely that India will get additional foreign reactors – at least in the near future—if this issue is not resolved. Therefore unless the issue of Sec. 17(b)—the Right of Recourse- is resolved there is no likelihood of any reactor being imported by India from any supplier.

The second impediment arising from this section is the Indian ratification of the Convention on Supplementary Convention (CSC). The CSC was signed by India on October 27, 2010 and India had promised, during President Obama's visit to India in 2011, that it would ratify the same before the end of 2011. However, it has not so far ratified the CSC. The US had contended that the Indian act was not in conformity with the requirements of the CSC and that India should consult with the IAEA on this. However, India had contended, quite correctly, that the only requirement for signing the CSC was that the concerned "State declares that its national law complies with the provisions of the Annex to this Convention."

In an earlier decision in respect of another dispute the International Court of Justice (ICJ), in its advisory opinion, had held that:

> Court considers that signature constitutes a first step to participation in the Convention. It is evident that without ratification, signature does not make the signatory State a party to the Convention...Until this ratification is made, the objection of a signatory State can therefore not have an immediate legal effect.

Therefore, the US would have no ground to either object to India's liability act or appeal against it. However, as soon as India deposits its ratification with the depositary of CSC—namely the IAEA—it would also have to provide the depositary with a copy, in one of the official languages of the United Nations, of the provisions of its national law and copies of such provisions shall be circulated by the Depositary to all other Contracting Parties.

Hence, once India ratifies the treaty and deposits its ratification to IAEA, the US can legally object to India's liability law on the ground that it is not in conformity with the provisions of CSC. At that point India cannot justify its signature/ratification merely on the basis that the CSC only requires that the "State declares that its national law complies with the provisions of the Annex to this Convention." As the ICJ had remarked in the earlier advisory opinion cited:

> It has nevertheless been argued that any State entitled to become party to the Genocide Convention may do so while making any reservation it chooses by virtue of its sovereignty. The Court cannot share this view. It is obvious that so extreme an application of the idea of State sovereignty could lead to a complete disregard of the object and purpose of the Convention.

In the present case too it is obvious that so extreme an application of the state's self certification would imply a complete disregard of the object and purpose of the CSC. Therefore, after India ratifies the CSC, the US objection to the Indian liability Act would have legal force and at that point any dispute between India and the US will have to be settled as per the procedures for dispute settlement laid out in the CSC. It is very likely that the US objection would be upheld and India's liability act found to be inconsistent with the provisions of the CSC. That, of course, would not be of any consequence. First of all very few countries have signed the CSC and it is yet to come into force. Secondly it seems very unlikely that the CSC would enter into force at any time in the near future. Finally, there is no compulsion that any country, including India, has to be a party to the CSC in order to engage in international trade. However, that would leave open the issue of foreign suppliers boycotting India because of its liability provisions.

(ii) Domestic

The domestic implications of Sec. 17(b) are equally, if not more, serious. So far the Indian nuclear programme has been dependent on Indian nuclear suppliers for various equipment, components and systems for the indigenous PHWR (Pressurised Heavy Water Reactor) programme. India has already built 16 such reactors with varying capacities, with four more under construction now and plans were afoot to begin construction of 10 more such plants during the XII Plan period. All this has been possible because of the development of an indigenous nuclear industry in the face of the international denial of nuclear technology to India in the past. One of the reasons why Indian industry was willing to invest in such a manufacturing capability and capacity was the liability regime—till the passage of the Indian Nuclear Liability Act. Till then the General Conditions of Contract for Supply of Indigenous Stores explicitly stated that, "6.7.7 The Purchaser shall indemnify and hold harmless the Contractor in respect of Third Party life and Property damage claims arising out of nuclear event at Purchaser's Site."

However, after the passage of the Indian Civil Liability for Nuclear damage Act, 2010, Indian suppliers too have become reluctant to supply equipment, components etc for the planned indigenous nuclear power plants and it seems very likely that the indigenous PHWR programme too will come to standstill because of Sec. 17(b) of the Liability Act, as the NPCIL has been unable to conclude any fresh contract for the supply of items for the planned power plants. Therefore, unless some remedial measures are taken, the Indian civilian nuclear power programme has reached a dead end and there will be no future for a civil nuclear power programme in India.

Possible Solutions to the Issues Arising out of Sec. 17(b)

(i) Amending the Act to remove Sec. 17(b): This is most unlikely given the controversy surrounding this section and the intense debate that took place in the parliament prior to its passage. This remedy can be ruled out.

(ii) Suitably restrict the disadvantages of Sec. 17(b) through a well drafted set of rules for the enforcement of the Civil Liability for Nuclear Damage Act.

The Government issued the Civil Liability for Nuclear Damage Rules, 2011 in November 2011, restricting the scope of application of Sec. 17(b). The main objection of nuclear suppliers to Sec. 17 (b) was that it imposed a potential liability, through the right of recourse, on the suppliers for an indeterminate amount for an indeterminate time lasting through the life of the plant. It did so in two ways. First, it restricted the right of recourse for the duration of the initial licence issued under the Atomic Energy (Radiation Protection) Rules, 2004 or the product liability period, whichever is longer. Under these rules, the initial licencing period is only five years. Hence under the new rules, the liability of the suppliers on account of the right of recourse will be only five years and not for any indeterminate period. Second, it clarified that that the Operator's claim under the right of recourse shall in no case exceed the actual amount of compensation paid by him up to the date of filing such a claim. Since the Liability Act limited the Operator's liability to a maximum amount to be set by the government—currently 300 million SDRs, the government also addressed the issue of indeterminate liability on the part of the supplier.

So under the current rules, the right of recourse is valid for only five years and that too, only up to a limit of 300 million SDRs. Hopefully these two major concessions—the maximum that can be granted by the government—will encourage both foreign and domestic suppliers of nuclear equipment, systems and components to enter into commercial agreements with NPCIL for supply of such items.

(iii) A third solution would be to encourage private operators to enter the field of nuclear power generation. Private operators will be free to forego the right of recourse based purely on cost/benefit analysis and may feel that it makes sense to forego such a right of recourse rather than be denied any supply at all—considering the extremely low probability of a major nuclear incident involving substantial claims.

(iv) The last solution would be for the suppliers to take out a liability insurance for the maximum amount for which they would be liable i.e. 300 million SDR, for a period of five years. The annual premium for such coverage is unlikely to exceed one million SDR per year. Such insurance can be taken either by the supplier himself and the resulting premium payments frontloaded on the contract or by the operator on behalf of the supplier the premium being paid by the operator.

(v) A petition on the constitutional validity of the Nuclear Liability Act, is at present, pending before the Supreme Court of India. If the Supreme Court in its wisdom strikes down the act as being unconstitutional then the status quo ante will be restored and then perhaps the NPCIL can sign its contracts with domestic and foreign suppliers in line with its earlier practice.

NOTES & REFERENCES

1. International Atomic Energy Agency, "PRIS: The Database on Nuclear Power Reactors", URL: http://www.iaea.org/pris/, accessed on July 25, 2013.
2. "International Conventions and Legal Agreements", *International Atomic Energy Agency*, URL: http://www.iaea.org/Publications/Documents/Conventions/liability.html, accessed on July 26, 2013.
3. "Vienna Convention on Civil Liability for Nuclear Damage, *International Atomic Energy Agency*, URL: http://www.iaea.org/Publications/Documents/Infcircs/1996/inf500.shtml, accessed on July 28, 2013; "Protocol to Amend the Vienna Convention on Civil Liability for Nuclear Damage", URL: http://www.iaea.org/Publications/Documents/Infcircs/1998/infcirc566.pdf, accessed on July 28, 2013, and, Convention on Third Party Liability in the Field of Nuclear Energy of 29th July 1960, as amended by the Additional Protocol of 28th January 1964 and by the Protocol of 16th November 1982, URL: http://www.oecd-nea.org/law/nlparis_conv.html, accessed on July 30, 2013.
4. Nuclear Energy Agency, "Brussels Supplementary Convention", URL: http://www.oecd-nea.org/law/brussels-supplementary-convention.html, accessed on July 27, 2013.
5. See, Vienna convention.
6. Convention on Supplementary Compensation for Nuclear Damage, *International Atomic Energy Agency*, URL: http://www.iaea.org/Publications/Documents/Conventions/supcomp.html, accessed on August 05, 2013.
7. Ibid.
8. Ibid.
9. Department of Atomic Energy (1962), "The Atomic Energy Act, 1962, No. 33 of 1962", September 15, 1962, URL: http://dae.nic.in/?q=node/153, accessed on August 06, 2013.
10. "Lok Sabha passes Nuclear Liability Bill, *Press Trust of India,* August 25, 2010, URL: http://articles.timesofindia.indiatimes.com/2010-08-25/india/28310404_1_nuclear-liability-bill-defects-or-sub-standard-services-latent-defects-or-sub-standard, accessed on August 10, 2013
11. A two-member committee comprising S. Rajagopal of the National Institute of Advanced Studies (NIAS), Bangalore, and V.B. Coutinho, Professor at Government Law College, Bangalore, was set-up in 2000 to evolve a suitable liability regime based on legislations in other countries. Their report, which was submitted in November 2001, however has not been made public.
12. Ministry of Law and Justice (2010), "The Civil Liability For Nuclear Damage Act, 2010", September 22, 2010, URL:http://lawmin.nic.in/ld/regionallanguages/THE%20CIVIL%20LIABILITY%20OF%20NUCLEAR%20DAMAGE%20ACT,2010.%20(38%20OF2010).pdf, accessed on August 12, 2013.
13. Ibid.
14. Rajya Sabha Starred Question No.72, "*New Nuclear Power Reactors in the Country*", Government of India, Department of Atomic Energy, URL: http://dae.nic.in/writereaddata/parl/mansoon2013/rssq72.pdf, accessed on August 13, 2013.
15. Rajya Sabha, Unstarred Question No: 2139, "*Liability In Case of Accidents*", Government of India, Department of Atomic Energy, URL: http://dae.nic.in/writereaddata/rsus150410.pdf, accessed on August 15, 2013.
16. Liu Jing and Michael Faure (2012), "Compensating Nuclear Damage in China", *Washington University Global Studies Law Review*, Vol. 11 (4), 2012, URL: http://digitalcommons.law.wustl.edu/globalstudies/vol11/iss4/2, accessed on August 14, 2013.
17. "Agreement between the Government of the French Republic and the Government of the Russian Federation on Third Party Liability for Nuclear Damage Caused in Connection with Deliveries from the French Republic for Nuclear Installations in the Russian Federation",

OECD, URL: http://www.oecd-nea.org/law/nlb/Nlb-66/075-080.pdf, accessed on August 15, 201

18. Ibid, Russia and France Nuclear Cooperation.
19. J. Venkatesan (2013), "Opinion on liability waiver based on legality, says AG", *The Hindu*, September 22, 2013, URL: http://www.thehindu.com/todays-paper/tp-national/opinion-on-liability-waiver-based-on-legality-says-ag/article5155467.ece, accessed on September 23, 2013.

Conclusions

Arvind Gupta

The foregoing chapters provide assessment of the various challenges and prospects for India's nuclear energy programme in the aftermath of Fukushima crisis in Japan. The Fukushima crisis has intensified the public opposition to new nuclear energy projects in India. The protests from local communities and certain civil groups at many new nuclear sites have raised some pertinent concerns over India's nuclear energy programme. This study has earnestly strived to evaluate some of the concerns vis-à-vis the overall performance of India's nuclear energy programme over the last few decades. The study highlights that while it may not be possible to address ideologically motivated opposition to country's nuclear programme, what merits serious discussion after Fukushima crisis is the safety of nuclear power reactors and the regulatory framework that oversees it. But more importantly, any discussion on desirability of nuclear energy must take into account India's growing energy demands and the imperatives of providing clean energy at competitive tariffs.

The disaster at Fukushima dai-ichi nuclear station in Japan has generated varied responses across the world. While Germany has decided to phase out its nuclear reactors by 2022, countries like India and China are clearly moving ahead with their respective nuclear energy expansion plans. The primary driver of India's nuclear energy expansion is the immediate need to bridge the massive energy demand and supply shortfall. India currently, suffers from huge energy deficit with almost 600 million people without full or partial access to electricity. India has not been able to provide energy access to many of its citizens which adversely impacted their quality of lives and overall economic development. The per capita consumption of electricity in the country has increased from 15.6 KWh/per year units in 1950 to about 766

KWh/per year units during the year 2009-10. By 2012, it was 778 KWh/per year units. This is considerably low compared to the per capita energy consumption in the developed countries. In order to sustain the current economic growth rate and ensure energy security, India will have to increase the supply and consumption of energy.

At present, India's energy basket is primarily fossil fuel centric (coal, oil and gas). Indian government's current policy is to increase the share of non-polluting energy sources such as renewable and nuclear. This is not without a reason. India's overall installed power generation capacity currently stood at 2,25,793.10 MW (as on June 30, 2013) comprising of 68% fossil sources, 18% hydro, 12% renewable sources and 2.12% nuclear. Power generation in the country is likely to remain fossil fuel based in the decades to come. While the share of nuclear energy currently, is mere 2.12%, it will steadily increase with the commissioning of new nuclear power projects in the near future.

Currently, there are 21 nuclear reactors in operation with an installed capacity of 5,780 MW and seven reactors of 5300 MW capacity are under construction. The government of India's Integrated Energy Policy in 2006, had drawn up an ambitious plan to raise the nuclear power generation capacity to 20 GW by 2020, 63 GW by 2032 and 275 GW by 2052 and thereby, increase the overall share of nuclear energy in country's energy mix to about 25 per cent by 2052 from its present share of less than 3 per cent. The targets set in the Integrated Energy Policy, however, proved to be far from achievable despite the exemption from the Nuclear Suppliers Group enabling nuclear commerce with India. Subsequently, in 2013, the DAE revised the targets wherein the installed nuclear energy capacity in India is expected to reach 10,080 MW by 2017, on the progressive completion of seven reactors under construction. It has set a target of generating 27,480 MW by 2023-24. Along with renewable energy sources, nuclear energy is supported as a non-polluting source of energy for environmental-friendly economic development and growth. Any additional nuclear-based power generation capacity would displace thermal power generation (Coal, Oil and Gas) and thereby, reduce carbon emissions and help mitigate climate change impacts. Indian policymakers have therefore, been committed towards increasing the share of nuclear power generation as a clean source of energy and achieving the set targets.

Nuclear energy nevertheless, suffered a decline in public acceptance in the wake of Fukushima nuclear crisis in Japan. The public concerns regarding nuclear energy after the Fukushima Dai-ichi accident, are mainly focused on the safety of the nuclear reactors, nuclear emergency management, impact on environment and population in the vicinity of the nuclear power plants, etc. The misinformation campaigns of certain civil society groups have partly stoked fears and opposition to nuclear energy. After Fukushima crisis, the NPCIL and AERB carried out extensive safety reviews of all the nuclear power reactors in India. These safety reviews have highlighted various strengths in the Indian reactor designs to withstand certain internal as well as external shocks. The safety reviews recommended several additional safety measures to safeguard nuclear reactors from external events such as tsunamis and cyclones. The Pressurised Heavy Water Reactor, (PHWR) which is a workhorse of India's indigenous nuclear energy programme, has been a great technological success for India's reactor designers and enjoys immense confidence of the Indian operators and regulators.

Post Fukushima crisis, the Indian PHWRs are specially evaluated for their ability to carry out core cooling under prolonged condition of total loss of power from reactors as well as the back-up systems. The AERB safety review have made detailed observations and recommendations regarding the emergency core cooling systems (ECCS) in the Indian PHWRs, which have been implemented by the NPCIL. The Indian PHWRs include multi-layered safety systems and redundancies to check possible malfunctions and failures of components as part of defence-in-depth approach. In addition, safety analysis exercise has been carried out through deterministic as well as the probabilistic safety analysis to ensure that internal and external events do not adversely increase the risk to the reactors. Also, the periodic safety reviews are regularly conducted during the operating life of the reactors.

The environmental impacts of NPPs have been the central concern for various civil society groups in India. The campaigns of these groups have propelled grass-root level opposition at many new nuclear energy sites in India. The DAE and NPCIL are mainly entrusted with the task of minimising adverse impact of the NPPs on the environment and surrounding population. The AERB approves the environment and radiation protection programmes at all nuclear installations. It has enacted stringent regulatory framework in the form of various codes and guidelines to ensure environmental and radiation protection during the siting, construction, operation and de-commissioning stages of NPPs. Additionally there are various statutory requirements which evaluate possible environmental and ecological impacts due to construction and operation of NPPs. The Ministry of Environment and Forests (MoEF) have put in place a mandatory requirement of environmental clearance for setting up nuclear installations (NPPs, Mines, Reprocessing Plants etc) in India. The environmental clearance for nuclear installations is granted by the MoEF only on satisfactory completion of comprehensive Environmental Impact Assessment (EIA) and Public Consultations. Additionally, the environmental clearance is granted on the basis of specific conditions, which the project authorities need to comply with. The MoEF can veto the clearance on grounds of ill-suitability of the project or non-compliance with the environmental protection rules.

During the operational stages of NPPs, radiation monitoring is regularly carried out as per the AERB guidelines by the network of Environment Survey Laboratories (ESLs), established at each NPP. The ESLs collect the comprehensive baseline data before the start of nuclear plants and thereafter, conduct regular monitoring during the operational stages of the NPPs. The ESLs examine various environmental matrices such as water, soil, food crops, goat thyroid, etc. to ensure that there is no build up of radioactivity in the environment over and above the pre-operational natural levels. Monitoring activity is carried out throughout the plant's operational life to analyse long term trends of radiological releases.

The environmental radiation monitoring programme at Indian NPPs has not shown any measurable increase in the pre-operational radiation levels. It should be noted that the natural sources of radiation by far, are the largest source of radiation exposure for the human population on earth. The doses received by both radiation and non-radiation workers at the nuclear plant are far lower than the prescribed regulatory limits and pose no serious health hazards to the employees and their families. The instances of cancer among the nuclear plant workers are also comparable with that of national cancer average. One can thus, reasonably surmise that radioactive releases from nuclear power plants and nuclear fuel cycle facilities in India, do not pose any serious health hazards to the public health and environment.

Apart from the reactor safety, doubts have been raised about the effectiveness of nuclear emergency plans and preparedness to deal with nuclear accidents in India. There has been a general lament about India's overall preparedness to deal with any major ecological disaster, let alone large-scale radioactive contamination. The civil society groups have overtly criticised DAE's technical as well as administrative preparedness to deal with Fukushima type nuclear disaster. In the light of the above concerns, post-Fukushima, the AERB has directed the NPCIL to review its capabilities to deal with nuclear emergencies. A comprehensive nuclear emergency preparedness plan is a mandatory AERB licensing requirement for setting up nuclear power plants. The AERB has laid down specific guidelines and codes for management of the nuclear emergency preparedness at NPPs.

Additionally, the guidelines enacted by the National Disaster Management Authority (NDMA), mandates the district and state administration hosting NPPs to develop requisite capabilities to deal with the off-site nuclear emergency. The district and state administration receives overall guidance and technical support from the DAE and NDMA for ensuring off-site emergency preparedness. This include the training and capacity building at the level of district and state administration, particularly the police, transport, health, and other agencies to carry out evacuation and to protect public from excess radiation exposure. The DAE has also developed various early radioactive monitoring and detection systems such as Indian Environmental Radiation Monitoring System (IERMON) with data transfer facilities, GIS based Emergency Response System and Indian Real Time Online Decision Support System (IRODOS). Also, an elaborate network of 20 Emergency Response Centres (ERC) spread all over the country including the NPPs sites with nodal ERC at the Bhabha Atomic Research Centre, has been set up. These ERCs are fully equipped in terms of instruments and trained personnel to deal with various types of nuclear emergencies. Post Fukushima crisis, the NDMA is setting up additional ERCs, while expanding the existing ones. The increasing technical competence and situational awareness of the NDMA had been witnessed during the successful evacuation of at least 1 million people from the coastal districts of Odisha and Andhra Pradesh within 12 hours of early warning before Cyclone Philian hit the East Indian coast in October 2013. A comprehensive preparedness is thus, ensured to deal with any disaster with large-scale radioactive releases as part of the overall nuclear safety objective. So far, there has not been any incident at Indian NPPs leading to off-site emergency response.

Finally, there is considerable anxiety amongst the civil society groups about the safe disposal of radioactive waste generated from NPPs. Given the widespread fears, apprehensions and anxieties around the radioactive waste disposal from NPPs; the safe and effective management of radioactive waste becomes crucial for public acceptance of nuclear energy. In order to minimise the waste generated from nuclear operations, India has adopted a reprocess and recycle approach. The safe handling of radioactive waste is ensured through AERB' various codes and safety guidelines for radioactive waste disposal. The AERB guidelines are in consonance with the international principles of radiation protection formulated by the International Council on Radioactive Protection (ICRP). India's approach to radioactive waste management is primarily aimed at minimising waste volumes generated from fission process. Through reprocessing and partitioning techniques, the overall quantity of nuclear waste is reduced significantly. India's waste management approach so far, has reasonably safe record of the waste handling for more than four decades and is comparable with international practices.

The DAE and NPCIL have ensured utmost compliance with the various AERB safety codes and guidelines on plant siting, reactor design, construction, commissioning, operation & maintenance and decommissioning. At all these stages of nuclear power operations, the AERB approvals are critical. The AERB plays not only crucial role in the developing the safety codes and standards for Indian nuclear reactors but also ensures their timely enforcement and compliance. However, post Fukushima nuclear accident, various civil society groups and independent experts have questioned the AERB's non-autonomous status and its impact on AERB's regulatory mandate. While the AERB's current status has never compromised it's regulatory mandate; in view of the current international practice of separating the legislative framework related to promotional and regulatory aspects of nuclear energy and to further enhance the legal status of AERB, the Government of India has introduced the 'Nuclear Safety Regulatory Authority Bill' (NSRA Bill) in the Parliament in 2011. The bill is currently pending in the parliament and should be expeditiously passed. The NSRA would require a large number of independent experts to enforce regulation, which currently does not exist in India outside of the DAE. The NSRA would therefore, have to promote academic programmes in nuclear engineering in Indian Universities and institutions such as the Indian Institutes of Technology, National Institute of Technology, etc. Apart from nuclear engineering, the NSRA can collaborate with independent institutions in civil engineering, seismology, radiation dosimetry, etc. for drawing independent expertise. The regulation enforced by the independent regulators can significantly contribute to restoring public faith in country's nuclear regulatory framework and boost confidence in nuclear safety.

In order to improve the public interface with nuclear energy, the DAE has launched series of outreach programmes at existing and new nuclear project sites. Through public interactions, the DAE seeks to improve public knowledge on technical issues involved in nuclear energy. It is important to emphasize that the probability of large-scale accidents at nuclear power plants is infinitesimally low. However, the NPCIL is mandated to conduct a proper risk assessment using various qualitative and quantitative methods for its operating nuclear power plants. The modern nuclear reactors are designed to withstand a 9/11 style plane crash into a reactor containment. The light water reactor commissioned at Kudankulam plant in July 2013, includes all the advance safety features that protect it from diverse internal as well as external shocks. The debate on safety of Kudankulam project has unfortunately overlooked these factors and the public opposition is propelled by certain misinformed campaigns.

The expansion of nuclear energy in India, currently, confronts various challenges in the form of opposition from the civil society groups and constraints imposed by the country's nuclear liability act. The prolonged persistence of such challenges could seriously hamper the DAE's ability to meet the current nuclear energy generation targets. While the chances of opposition from certain civil society groups to the nuclear power petering out in the near future seem unlikely, the NPCIL and DAE must expand their outreach programme for spreading awareness about the safety of its nuclear reactors. The whole-hearted public engagement can go a long way in improving public acceptance of nuclear power in India.

Contributors

Dr Arvind Gupta currently is the Director General of the Institute for Defence Studies and Analyses and Convenor, Indian Pugwash Society.

Dr K. D. Kapur is a ICSSR Fellow at the Institute for Defence Studies and Analyses and former Professor of Diplomacy, Jawaharlal Nehru University, New Delhi.

G. Balachandran is a Consulting Fellow at the Institute for Defence Studies and Analyses. He has a PhD in Economics and Computer Science from the University of Wisconsin.

Mr Sudhinder Thakur is Executive Director, Corporate Planning, Nuclear Power Corporation of India Limited, Mumbai

Shri P.K. Wattal is Director, Nuclear Recycle Group Bhabha Atomic Research Centre, Mumbai

Kapil Patil is a Research Assistant at the Indian Pugwash Society. He holds an MPhil in Disarmament and Diplomacy Studies from School of International Studies, Jawaharlal Nehru University, New Delhi.

Index